BIO-CULTURAL EVOLUTION

by the same author

- Developmental Anthropology
- जटिल समाज का मानवशास्त्र
  (Anthropology of Complex Society)
- भारतीय मानवशास्त्र
  (Indian Anthropology)
- भारतीय जनजातीय संस्कृति
  (Indian Tribal Culture)
- मानवशास्त्रीय सिद्धान्तः शास्त्रीय एवं आधुनिक
  (Anthropological Theory: Classical and Modern)
- मानवशास्त्रीय अनुसंधान विधि एवं तकनीक
  (Anthropological Research Method and Technique)
- History of Anthropological Thought (Jointly with **V.S. Upadhyaya**)

About the Author

Gaya Pandey (*b.* 1954) is senior faculty member of P.G. Department of Anthropology, Ranchi University, Ranchi. Since 1985 he is actively engaged in teaching and research work related to tribal studies, socio-cultural anthropology, developmental anthropology, medical anthropology and linguistic anthropology. His dozens of research articles have been published. He has also co-authored nine popular course books in different branches of anthropology.

Bio-Cultural Evolution

Gaya Pandey

CONCEPT PUBLISHING COMPANY PVT. LTD.
NEW DELHI-110 059

ISBN-13: 978-81-8069-705-0 (HB)

First Published 2010

Published and Printed by

Concept
Since 1974

Concept Publishing Company Pvt. Ltd.
Regd. Office:
A/15-16, Commercial Block, Mohan Garden
New Delhi-110059 (India)
Phones : 25351460, 25351794, *Fax* : 091-11-25357109
Email : publishing@conceptpub.com
Website: www.conceptpub.com

Editorial Office:
H-13, Bali Nagar, New Delhi-110 015, India.

Cataloging in Publication Data--*Courtesy:* D.K. Agencies (P) Ltd. <docinfo@dkagencies.com>

Pandey, Gaya, 1954-
Bio-cultural evolution / Gaya Pandey.
p. cm.
Includes bibliographical references (p.) and index.
ISBN 9788180697050

1. Human evolution. 2. Social evolution. 3. Physical anthropology. I. Title.

DDC 599.938 22

Dedicated to

Wife Smt. Indu Devi, Daughter–Sudha, Suman, Sarita & Anu, Son Shekhar and Grand son Kushagra Divyansha who have been my main source of encouragement in completing this book even in my period of recent illness and cure.

Preface

Bio-cultural evolution is a paper taught in undergraduate and postgraduate Anthropology Departments of different universities of our country. This paper provides the students and scholars of Anthropology with an occasion to know about the relation between biological evolution and cultural evolution. Without the knowledge of biological evolution, it is very difficult to know about cultural evolution. Man is master of animal kingdom because his physical features are superior to other animals. His superiority over other animal is also hidden in his ability to create culture. Thus, human evolution has both biological and cultural dimensions.

Although informations about biological evolution and cultural evolution appear in different books of Biological Anthropology and Pre-historic archaeological Anthropology, but in those books one does not find materials on bio-cultural evolution in integrated manner. Therefore, students offering this course as one of papers in P.G. Departments of Anthropology are facing troubles. Keeping their troubles in mind, attempt has been made to present materials systematically within one book through simple language on bio-cultural evolution. However, it is very difficult to present biological evolution and cultural evolution in a single book because several books have been written on biological and cultural evolution separately. But attempt to understand biological evolution and cultural evolution together is really a tedious as well as praiseworthy task done in this book. The real evaluators of any book are readers and scholars. If they are satisfied, it will be the greatest achievement of the book and author. I request readers and scholars to make me aware of weakness of the book so that attempt will be made to improve the book in future.

Data and informations presented in this book have been gathered through library work. Hence, they do not form any new contribution of the author. The works of different schools have been reviewed and reproduced sometimes in toto to make the understanding clear for the benefit of students.

This book could achieve its success with the help extended by different persons at different level. Hence, it is my moral duty to express gratitude to them. First of all my sincere gratitude goes to all teachers and writers whose works and creations on the subject have been beneficial for me in composing this book. I am deeply indebted to my colleagues of the Department of Anthropology, Ranchi University, Ranchi, who not only gave valuable suggestions but encouraged a lot in accomplishing this task.

I am thankful to my wife Indu Pandey, daughters Sudha, Suman, Sarita and Anu and my only son Shekhar who cooperated me much in completion of this book. I am also thankful to my only grandson Kushagra Divyansha with whom I played very much whenever I got tired off composing the book.

I am thankful to Shri Ashok Kumar Mittal and his team, Concept Publishing Company Pvt. Ltd., New Delhi for getting the handwritten manuscript of the book published with great care.

One cannot claim to be perfect in expressions. Therefore, readers might excuse me for mistakes which might still be there inspite of our best efforts.

Gaya Pandey
P.G. Department of Anthropology
Ranchi University
Ranchi-834 008
Jharkhand

Contents

1
Introduction

Evolution means change in the direction of progress. In other words, it is progress oriented change. As a process, it is stage after stage progress in unilineal direction. The direction of evolution is always from simple to complex, from homogeneity to heterogeneity and from similarity to dissimilarity. Evolution reveals that something changes, while remaining partially the same. Evolution is associated most closely with biology, but it can be applied to any natural and social sciences attempting to reconstruct the past. Lamarck's and Darwin's theories were evolutionary. Similarly Marxist theory of dialectical materialism and enlightenment schemes of universal history were evolutionary. So was the classical cultural evolutionism of 19th century.

Bio-cultural evolution is associated with biological and cultural dimensions of human evolution. Human evolution has two significant dimensions simultaneously. On one hand man is a biological animal and on the other hand man is a cultural animal. Man is the master of animal kingdom. Being master of animal kingdom, he is most evolved animal of the earth. At the same time, man is the creator of culture. This feature of man distinguishes him from other animals. As man is the most evolved animal biologically and culturally, so human evolution has biological and cultural dimensions. Biologically human evolution has evolutionary origins and diversities of species Homo sapiens. Culturally human evolution also has evolutionary origins and diversities of cultural forms in different periods.

Relation Between Biological and Cultural Evolution

Since man is the most evolved animal of the earth biologically and he is the creator of culture, so there is close relation between biological and cultural evolution.

A good number of anthropologists hold view that cultural evolution is an extension of biological evolution because hominid was the minimum condition for the evolution of culture. Non-human primates were not the creators of culture. So hominid was the minimum condition related to the evolution of culture.

Changes in Man Responsible for the Evolution of Culture

In order to understand the relation between biological evolution and cultural evolution, it is necessary to know the physical changes that have occurred in man.

Physical changes in man responsible for creation of culture are as follows:

1. Erect Posture
2. Freely Opposable Thumbs
3. Stereoscopic and Chromatic Vision
4. Intelligent Brain
5. Creation of Symbols

1. *Erect Posture*: Man is the only animal who has efficiency to stand vertically followed by bipedal walk. The other animals follow quadrupedal walk. On account of this physical feature man uses his hands in performing work and creating culture. Other animals on account of their quadrupedal walk are not in a position to do work and create culture.
2. *Freely Opposable Thumbs*: Human beings have the efficiency to move hands freely in any direction. His thumbs are freely opposable by which he can catch anything easily. His efficiency of freely movable hands makes him competent to easily handle tools, implementing, machines, art, architecture, painting, drawing, etc. Other animals lack this physical feature, so they are not in a position to create culture.
3. *Stereoscopic and Chromatic Vision*: Man possesses stereoscopic and chromatic vision. On account of this physical feature man is able to make observation on natural and other events. On the basis of his observation he is in a

position to make various kinds of inventions and discoveries. As an inventor and discoverer, he is able to create culture.

4. *Intelligent Brain*: Human brain is more developed, intelligent and logical. His brain is certainly more thoughtful and active in comparison to other animals. The cranial capacity of human brain is roughly four times more than anthropoid apes. Due to intelligent brain, man is able to think, plan, express ideas and find out cause-effect relationship. All discoveries and inventions are product of intelligent brain.
   Man has such a logical brain which makes difference between dirty water, clean water and holy water. He makes difference between wife, mother, daughter and sister. Man knows the importance of magic, religion and science. Man makes exchange and borrowing. Thus intelligent brain is an important physical feature for the creation of culture.
5. *Creation of Symbols*: The efficiency of making symbols is unique physical characteristic of man. Other animals do not have this biological and cultural features. Other animals also produce sound symbols but they are very few in number. Man has enormous efficiency to create symbols. By altering the sounds, he can make varieties of symbols. He can make a number of verbal and non-verbal symbols by which he can take part in speech community. Each symbols of man has meaning. It is meaning which decodes the symbol for the users.

Considering the above physical features of man in relation to the creation of culture, a number of scholars consider cultural evolution as an extension of biological evolution. Because without changes in such physical features as erect posture, freely opposable thumbs, stereoscopic and chromatic vision, intelligent brain and creation of symbols man would not have been in a position to create and evolve culture. Thus, hominid was minimum condition for the creation and evolution of culture. In this sense, it is true that cultural evolution is an extension of biological evolution. Because without biological evolution in the form of hominid, the evolution of culture was impossible. It is true that there is relationship between culture and biology but cultural evolutionism does not represent an application of Darwin's biological ideas to the realm of culture. Cultural evolutionists were far more interested in ethnography, archaeology and expanded view of universal history than in Darwin's theory of evolution by natural selection. It would be inaccurate to label cultural evolutionists as social or cultural Darwinists.

Differences Between Biological and Cultural Evolution

It is true that there is relation between biological evolution and cultural evolution; because man is both biological as well as cultural animal. But this relation does not tell us that both biological and cultural evolution are same. There is profound difference between the principles of biological evolution and cultural evolution. The nature of evolutionary scheme and development process differ profoundly in biology and culture. The main points of differences between biological evolution and cultural evolution are as follows:

1. In biological evolution, it is assumed that all forms are genetically related and their development is essentially divergent or convergent. Parallels such as swimming, flying, walking, etc. are superficial and fairly uncommon. They are genetically considered to be instances of convergent evolution rather than parallels.
   In cultural evolution, it is assumed that cultural patterns in different parts of the world are genetically unrelated yet pass through the same parallel sequences. The divergent trends, caused by local environment, areas, sub-areas which do not follow universal sequences are attributed only to secondary significance. This was the basic assumption of not only 19th century classical cultural evolutionists, but neo-cultural evolutionists like White and Childe also escaped from the awkward facts of cultural divergence and local variations.
2. According to V.G. Childe, the nature of biological evolution is divergent but nature of cultural evolution is convergent.
3. According to Kroeber, the process of cultural evolution is additive and accumulative but the process of organic evolution is substitutive.
4. In biological evolution all human beings are descendants of same species Homo sapiens. In cultural evolution, the descendents of same species homo sapiens are creators of different kinds of culture.
5. Give and take of biological traits take place through

reproduction and heredity. But give and take of cultural traits take place through contact, communication, mutual borrowing, diffusion, migration, etc.

6. In biological evolution adaptation with environment takes place through mutation but in cultural evolution adaptation with environment occurs with the help of change in human behaviour.
7. Biological traits are genetic heritage, but cultural traits are social heritage.
8. Cultural traits are acquired easily from any group, but biological traits are not acquired easily from any groups.

Areas of Study

In order to understand the bio-cultural evolution, it is necessary to know the evolution of Homo sapiens and culture in different periods of time. If we look at the background of biological and cultural evolution, we come to know that a proper understanding of bio-cultural evolution requires knowledge about the following:

- Origin of earth, life and geological time scale.
- Position of man in animal kingdom
- Different theories of origin of life
- Principles of evolution
- Evidences of evolution
- Hominisation process
- Pleistocene and its main features
- Pre-historic culture and cultural periods
- Australopithecine stage with lower Palaeolithic pebble tool culture
- Pithecanthropous stage associated with lower Palaeolithic handaxe-cleaver tradition
- Neanderthal stage with middle Palaeolithic cultures
- Modern man stage with upper Palaeolithic culture
- Mesolithic culture and Coummunities
- Neolithic culture and Communities
- Pre-historic societies and social organisations
- Aesthetic point of culture as revealed from the pre-historic art
- Economic organisation of pre-historic societies
- Cultural perspectives of pre-historic societies

2

Birth of Earth and Evolution of Life

Geological researchers have established that our earth is nearly 4.5 billion years old. Today we all know that the shape of our earth is like orange and it moves around the sun. Its diameter is 8,000 miles and circle is 25,000 miles. These facts have been established today by scientific researches. But it was very natural for early man to think on the basis of his experiences that the earth was not round rather it was flat. The sun and moon moved around the earth. Such opinions were prevalent in Babylon and Egypt for centuries in spite of several researches made in the field of astronomy. In India, Arya Bhatta (b. 479 A.D.) established the theory that sun was stable and it was earth which moved around the sun. According to him the circle of earth is 24,835 miles. But unfortunately his theory could not find place even in our own country among his contemporaries. In Europe, it was Nicholas Copernicus (1473-1543) who established that the earth moved around the sun and that the earth moved on its own axis. Since then our knowledge about earth and creation has increased. Now we know that earth is a planet. It is one of members of solar system. It is situated at a distance of 9 miles to 9.5 miles from the sun. Other planets of earth are situated at hundred crores miles. Our solar system is one of solar systems of Akashganga. The Akashganga itself is one of Akashgangas of the creation. There are some planets of the creation whose light moves with a speed of 1,86,000 miles. The light of some planets does not reach on earth even in one billion years. In this creation our earth is like a drop in the ocean.

Birth of Earth: The concept about the birth of earth among early man was very confusing. In this field, too, no scholars of the world other than India had reached near the truth. In Europe such belief was prevalent till 18th century A.D. that God had created the earth in 4004 B.C., i.e. nearly 6000 years before today. First of all he created sky and then plants, animals and men. He had created sky, plant, animal and man only in six days. This confusing opinion was based on Bible of Jews. In Quran, too, the same opinion one finds regarding the origin of earth, plant, animal and man. One also finds similar opinion in religious book of Parsis called Awesta. But in present day it has been established on the basis of astronomical, geological, Palaeobotanical and Palaeontological researches that earth and other planets were parts of the sun. Nearly 4.5 billion years ago, when the earth and other planets were not in existence the size of sun was very very big. Suddenly an explosion took place in that very very big sun. It became possible due to dashing of another planet against the sun. On account of this natural phenomenon, vast waves were emitted from the sun. One of waves was separated from the sun and spread in the form of drops. The different spread drops assumed the shape of earth and other planets. Due to the force of attraction they began to move around the sun. Thus, our earth came into existence nearly 4.5 billion years before today.

Evolution of Life. It is very difficult to say as to when and how life originated on the earth. In ancient time such belief was prevalent that the God had created all plants and creatures simultaneously. They continued generation after generation. But today scientists mostly believe that life originated on the earth due to physical and chemical process. Due to physical and chemical processes an element called protoplasm was formed. The protoplasm became the base of life. Life developed in shallow water in presence of light in the form of acellular organism called protozoa. This microscopic organism was without bone, skin and hide. In course of time, change in external conditions brought changes in the organism. As a result of this multi-cellular organism called metazoan came into existence. This theory was propounded by Lamarck of France and Darwin and Alfred-Wallace in England. Besides evolution, a number of theories regarding the origin of life on the earth have been propounded. Let us have a discussion about these theories one by one.

Theories of Origin of Life

Different theories propounded by different scholars regarding the origin of life on earth are as follows:

1. Special creation theory
2. Spontaneous generation theory
3. Cosmozoic theory
4. Catastrophism theory
5. Chemical evolution theory
6. Evolutionary theory

1. *Special Creation Theory*. According to this theory life was created on the earth by super human power. Such type of opinions are prevalent in different religious scriptures. According to Bible, the entire world and life was created by super human power within six days. On first day, earth and heaven, on second day sky, on third day land and vegetations, on fourth day sun, moon and other planets, on fifth day bushes and aves and on sixth day man were created. From the twelveth rib of the man, a woman was created. This opinion was prevalent till mid of 18th century. In Hindu religion, Lord Brahma is regarded as the God of creation. He created all living beings. Living beings created in the past were in the same form as they exist today. This theory does not believe in evolutionism. But on the basis of scientific ground, this theory is not acceptable.
2. *Spontaneous Generation Theory*. According to this theory different forms of life emerged from the rotten organic substance spontaneously. Protozoa, bacteria, insects, flies, etc. originated due to decay of organic substance. This opinion occupied place till the middle of 16th century. But on the basis of scientific researchers, this theory was rejected. This theory was rejected by Luis Pasture, Reddy and Splajohny. Mr. Luis Pasture demonstrated an experiment. He left the boiled meat to decay in the same flask closed with cotton. Cotton was used to prevent the entrance of different bacteria under the flask with boiled meat and also

to make the air get entered in the flask. But the decay of meat could not originate life. On the basis of this scientific experiment, the theory of spontaneous generation was rejected.

3. *Cosmozoic Theory*. This theory postulates that bacteria, spores and seeds coming from different planets were responsible for creation of life on the earth. This theory was propounded by Liebig, Kelvin and Arrhenius in 1921 initially. But Francis Crick and Lerali Orgel propounded this theory strongly. Recent researches have shown that bacteria, spores and seeds coming from different planets on earth get destroyed in the way due to different types of radiation. Therefore, cosmozoic theory could not be accepted on this ground. This theory is also known as theory of directed panspermia.
4. *Catastrophism*. This theory was established by George Cuvier. According to him, different catastrophic events such as flood, volcanic eruption, submergence of earth in water, creation of mountains, turning of desert into seas, going down of sea water etc. were responsible for origin of different types of life. Catastrophic events had changed the geographical environment of the world to originate varieties of life on the earth in different environment.
5. *Chemical Evolution Theory*. This theory was propounded by famous scientist Operin in his book, 'Origin of life' (1924). He expressed the possibility of a biogenic reproduction from hydrocarbons, the simplest organic substance. According to this theory only carbons were present on earth in the earliest time. They were mixed in the water of sea. When these carbons were organised in special way, complex organic substances were created. From the complex compound such formed, different forms of life originated on the earth. According to this view, protein was formed from organic substance. Protein could develop colloidal system which ultimately was responsible for origin of life on earth. S.L. Miller was pioneer of this theory. But this theory also seems to be doubtful.
6. *Evolutionary Theory*. The most accepted theory of origin of life on earth is evolutionary theory. The pioneers of this theory

> were Lamarck, Darwin and Alfred Wallace. According to evolutionary theory the off-springs of all creatures are like to their parents. Their physical traits and nature are not entirely similar to their parents. On the other hand, each creatures have to adapt themselves with natural conditions in order to survive. According to Darwin, the number of off-springs of each species is much more than nature can provide food to them. That is why there is struggle for existence in all kinds of species. In their struggle for existence, those parts of body, which are best suited to the changed environment, evolve and others become extinct. Such continuous change brings changes in the species of a creature. Thus, according to this theory only those creatures survive who adapt themselves with changed environment. Those who do not adapt with changed situation become extinct. This law of evolution is known as natural selection or the survival of the fittest. For example, we take an insect which lives in black dry place. Majority of off-springs of this insects are black or red and some are green. If the condition changes and the place becomes green. In this situation, the green coloured insects will be in better position to adapt with the changed condition in comparison to red coloured or black coloured insects. The green coloured insects can protect themselves from enemies by hiding in green leaves. As a result of this adaptation, the number of green coloured insects will increase and red or black coloured insects will decrease. In this context, it is important here to mention that it is not essential that each link of evolution could be connected to other links. One can find jump from one stage to another in course of evolution. It is also important to note here that it is not essential that lower forms could become extinct after the emergence of higher forms in course of evolution. Creatures of lower forms are also found existing along with creatures of higher forms.

Darwin implemented his evolutionary theory not only on animals, but on man, too. After that a number of scholars made corrections in this theory. The idea of Darwin that off-springs get all developed elements of parents was rejected by German scholar August Weismann. According to Weismann, there are two types of cells in animals—somatic and genetic. Changes in somatic cells do not bring any change in genetic cells. An off-spring can inherit only those traits of parents which are present in their germ plasm. In 1901, Dutch scholar Hugo Devries propounded the theory of mutation on the basis of Mendelian hereditary theory. According to him, mutation brings changes in the species. Evolutionary theory has been described in detail in next coming chapters.

Evidences of Evolution: Evolution begins from the birth of life. In the early time life appeared due to self-reproduction of some special organic substances. Scholars are of opinions that sign of life appeared from organic substances. From them creatures with simplest structure emerged. Again from simplest creatures highly complex and developed form of creatures appeared on the earth. The term evolution was used for the first time by Herbert Spencer to present different organisations continuously. Charles Darwin used this term to show changed relation between ancestors and descendants or changed biological characters. Regarding biological evolution we can say that:

1. Highly complex and advanced form of life has evolved from very simple form of life.
2. Biological evolution is a continuous process. All creatures have changed according to time and they will continue to change in future.
3. Creation of new species is highly complex, slow and time taking process. It takes place in several hundred thousand years. It cannot be observed directly.
4. There are certain evidences which help us in understanding the evolution.

Chief evidences of evolution are as follows:

1. Evidences from comparative anatomy
2. Evidences from taxonomy
3. Evidences from geographical distribution
4. Ecological evidences
5. Evidences from embryology

6. Evidences from palaeontology
7. Evidences from physiology and bio-chemistry
8. Evidences from genetics
9. Evidences from molecular biology

1. *Evidences from Comparative Anatomy*: Evidences from comparative anatomy has been proved helpful in understanding the evolution of different animals. Some important evidences are as follows:

   (i) *Homology and Analogy*: Some similarities in the anatomical structure of different animals appear but their functions are different. Their similarity is based on common ancestor. Such organs are called homologous organs. Animals related through common ancestry show homologous structure. Different advanced animals show homology in the structure of organs. For example one finds homologous in he structure of feet of bat, forearms of horse and forearms of man. But functions of these homologous organs in different animals are different. The homologous organs of bat, horse and man bear testimony that in the past these animals had a common ancestor from where they have evolved.

Contrary to the evidence of homology is the evidence of analogy. Evidences of analogy also suggest about the evolution of animals. Some animals show analogy in the structure of organs, but the same funçtion is performed by them. On the basis of function those organs look similar externally but they are different physically. Analogous organs would have been created to perform different functions in the early time. But in course of time, they began to perform similar function. For example, butterfly, aves and bat have flying organs. The flying organs of butterfly, aves and bat perform the same function, but flying organs have developed differently in them. They exhibit similar functions, but their structures are different anatomically. These evidences suggest that these animals possessed different ancestors. In this way one finds that analogous and homologous organs found in different evolved animals provide us evidences in support of the evolutionary theory.

(ii) *Vestigial Organs*: Vestigial organs bear testimony regarding the evolution of animals. Vestigial organs are those organs which are useless or unessential. Vestigial organs would have performed significant functions in ancestors, but they have became useless or unessential for the descendants. Vestigial organs are remains of developed organs. These organs used to perform significant functions in the ancestors, but after evolution they either became extinct or appeared as remains. For example, man and other animals show a number of instances of vestigial organs. The vermiform appendix in man is a vestigial organ. Sometimes, it becomes the cause of appendix disease in man. Appendix is found in developed form in other primates. In these primates, it helps in the digestion of coarse food materials. Similarly, the presence of three to five remains of tail vertebral in man suggest that the ancestors of man possessed tail. In place of these vertebral one finds presence of tail in other primates. In this way, vestigial organs present evidence about the evolution.

(iii) *Adaptive Modifications*: Adaptive modifications in the organs of animals belonging to same class present strong proof regarding the evolution. For example, the forearms of animals belonging to class mammalia show adaptive modifications. In the common scheme, the forearms of mammals show one humerus, one radius, one ulna, eight carpels of wrist arranged in two lines, five metal carpels in the structure of palm and three finger bones in each finger. But in spite of this similarity in the structure of forearms of mammals, these bones show adaptive modifications. According to functions in different times, they have assumed different forms in changed situations. Thus, adaptive modifications in some bone of animals belonging to same class prove that evolution has taken place.

2. *Evidences from Taxonomy*: There are nearly 10,00,000 species of animals and 2,50,000 species of plants in the world. These animals and plants have been classed on the basis of similarities and differences in animal kingdom and plant kingdom. It was Linnaeus Carolus (1707-1778), a Swedish biologist, who classified living thing in a hierarchy of taxonomic category, using a system of binomial nomenclature or two names for the category of species. In his book, system of nature Linnaeus introduced many of the taxonomic names such as kingdom, phyla, class, order, family, genus species, etc. These names are used in evolutionary biology today. Thus evidences from taxonomy also presents a strong proof in support of evolution.
3. *Evidences from Geographical Distribution*: Geographical distribution is one of causes for creating differences in the biological structure of animals. Even closely related animals exhibit physical differences living in different geographical areas. For example, the presence of platypus of class protheria and kangaroo of class marsupials found in Australia. This continent till Cretaceous geological period nearly 12.5 crores years ago was connected to Asia geographically. But due to entrance of torres strait, these two continents were isolated geographically. Due to this geographical isolations, many plants and animal species, though related at that time, were isolated. Now the direction of evolution in both continents took local adaptation in changed environment. But in spite of local modification, they exhibit 'similarities'. Thus evidences from geographical distribution are strong proof for the evolution.
4. *Ecological Evidences*: On the basis of scientific facts it has been established that ecology affects the evolution of life. It was Charles Darwin who had attempted to understand the impact of ecology on the evolution of life. While on tour Darwin came across varieties of species of plants and animals when his ship Beagle sailed around Atlantic and pacific costs of South America. When he visited Galapagos Island, he observed varieties of finches and tortoises differed slightly from one island to another. They also differed from the varieties

on South American mainland. Darwin came to the conclusion that such varieties at different places of the word would have evolved due to ecological conditions. One finds the influence of ecological factors on evolution of human species too. One finds differences in the species of man living in islands, desert, forest, hilly areas, sea coast and plain areas. Thus, evolution can be understood on the basis of ecological conditions.

5. *Evidences from Embryology*: Embryological evidences also suggest about the evolution. It was Ernest Haeckel who postulated biogenetic law or recapitulation theory. On the basis of this theory embryological evidences were taken as strong proof of evolution. On the basis of this law attempts were made to understand that embryological changes in all multicellular animals exhibit similarity. In multicellular animals one finds the beginning of life in unicellular zygote, several divisions of eggs, formation of blastula and second stage gastrula after some time. After these stages, different animals exhibit some differences. In early stage, the embryo of fish, salamander, birds and man exhibits similarity to a great extent. This indicates that these animals would have evolved from aquatic fishes.
6. *Evidences from Palaeontology*: Palaeontological evidences are strong proof regarding evolution. Fossils of plants and animals have yielded useful stages related to evolution. When fossils are arranged according to time sequence, they present strong proof of stages through which the evolution of animals and plants had taken place. Entire geological time scale has been formulated on the basis of fossils discovered from different stratum of rocks. Fossils have been helpful not only in establishing geological time scale, but they have also helped us in solving the problem of missing-link.
7. *Evidences from Physiology and Bio-chemistry*: Similarity and difference in phenotypic traits of animals is determined by different bio-chemical products and physiology. On the basis of this fact one can say that evolution is a bio-chemical phenomenon. Several facts related to physiological and biochemistry present strong proof regarding the evolution.

Some important facts in this regard are as follows:

Protoplasm Chemistry: All creatures, unicellular or multicellular, possess living matters in protoplasm. Biochemical composition of living matters found in protoplasm of unicellular and multicellular creatures exhibits similarity. Protoplasm is composed of proteins, carbohydrates, fats, water, etc. This clarifies that basic property of living substance remains maintained in the process of evolution. This is a strong proof related to evolution.

Chromosome Chemistry: Like protoplasm, one also finds similarity in the chemical composition of chromosomes of unicellular and multicellular creatures. The chromosomes of all living beings are composed of nucleoproteins. Nucleic acid DNA found in it is composed of proteins, histone and protemine. Globulin and RNA are also found in chromosomes. Similarity in chromosome chemistry of all living beings reveals that all creatures have evolved from the same ancestor.

Enzymes and Hormones: Some animals possess similar enzymes and hormones. The presence of common hormones in those animals reveals their common ancestry. For example, trypsin enzyme is found in the digestive tube of all creatures. Similarly, thyroid enzymes are found commonly in all vertebrates. This hormones can be substituted from creatures of one class to another. The commonly presence of these enzymes and hormones in living beings bears testimony about the common ancestor in early time.

Serological Test: Different serological tests also reveal proof related to evolution of animals. For example, if immunised blood serum of human blood is divided into five groups and annexed respectively with immunised blood serum of man, ape, old world monkey, new world monkey and lemur. Closeness of these animals can be shown on the basis of tests or study of concentration of precipitation. The precipitate of closely related animals show more concentration or thickness. Thus, such serological tests are helpful in showing the direction of evolution of animals.

8. *Evidences from Genetics*: Different facts related to genetics are also helpful in understanding the process and direction of evolution. Genetic differences are created through

recombination and mutation. Hybridisation creates recombination as a result of which new combination evolves. Mutation is an important factor for creation of genetic materials which bring changes. Thus, similarity and differences in physical characteristics are taken as proof of evolution.

9. *Evidences from Molecular Biology*. Molecular biological facts also present evidences in support of evolution. Important evidences in this regard are as follows:

Evolution of Amino-acid Sequences in Proteins

Evolution of amino-acid sequences in different proteins is helpful in understanding the process of evolution. For example, in mammals a protein named haemoglobin difference in an amino-acid shows distance of 70 lakhs from other animals. Some other proteins like insulin, cytochrome, etc. show more distance of time. From the distance of time period the rate of evolution can be known. But it is essential to calculate time distance of amino-acid in more than one protein to understand the evolution. For example, the haemoglobin of man and gorilla shows difference of only one amino-acid, though they are not closely related. Therefore, one should calculate amino-acid structure in more than one protein.

Evolution of Nucleotide Sequence in Proteins

Nucleic acids like DNA and RNA possess nucleotides. By studying the nucleotide sequences of these nucleic acids, evolutionary processes can be understood. In this method, hybridisation of DNA invitro of two animals is done and difference is calculated. Difference of DNA is calculated on the basis of thermal stability. Such type of experiment shows that evolutionary distance of cow is less to pig than sheep. Attempt is done to make phylogenetic tree by calculating nucleotide substitutions and nucleotide sequences in DNA.

Evolution of Life

It is still impossible to tell about the exact date of origin of life on the earth for the first time. Geological researches have revealed that it would have taken nearly two billions years of time to emerge life on the earth from the date of creation of earth. At the time of its birth the earth was a dangerous ball of gaseous fire. Gradually it became cold. Its upper layer first of all attained liquid state and then gradually turned into solid state and at the end turned into rocks. Water would have been present in those days on earth in the form of vapours but on being cooled, they began to fall as rain drops. Now lakes, rivers and seas of the earth got filled with water. Due to continuous attack of rains and winds rocks were broken into small parts and assumed the shape of soil on the earth. In this process nearly 2 billion years of time would have taken place. Nearly 2.5 billion years ago the condition of the earth became favourable for the emergence of some kind of life on it. The period of time which was spent from the origin of earth to the origin of oceans on the earth is called by the geologists as *Cosmic time*. There is no evidence to study this period. After this period, we have evidences in the form of sedimentary rocks to study about the origin of life on earth. The period of time from the origin of life on earth to present day is called as Geological time by geologist.

Sedimentary rocks are those pages of geological history by which one can study about the evolution of life on earth. Sedimentary rocks are formed by the layers of sediments brought by agencies like river, air and glaciers. When rocks are subjected to weathering and erosion, sediments are formed. These sediments are brought by the above agencies and deposited at a place in the form of layer. In due course, silica, clay, carbonate, iron and salt are deposited in the loose particles of sediments. By the process of welding and cementation, the layers of those sediments assume the form of rocks. These rocks are formed in the shape of layers. In the stable water, bigger particles are deposited first and then smaller particles are deposited. Thus, layer of bigger particles are below and finer smaller particles are above. As a result of this process, layer after layer goes on deposited. These are called sedimentary rocks. In the layers of sedimentary rocks fossils of animals, plants and cultures belonging to those periods are found buried. Signs of rains and foot prints are also found buried. Dating of layers of sedimentary rocks have been done by the scholars. On the basis of that study the historical reconstruction of human civilisation is done. Sedimentary rocks are of various kinds such as sand made rock, called sand stones, pebbles and gravel mixed rocks called conglomerate. Pieces of conglomerate are round or oval in shape because they are swept by river current for a long time.

The scientists have studied the remains of plants, animals and culture found in sedimentary rocks as fossils. On the basis of that study, the history of evolution of life has been presented under geological five chapters. These are as follows:

1. *Azoic Age or Archaeozoic Age*: The study of sedimentary rocks reveals that its earliest layer was 270 crore years old. In these layers, 190 crore years old layers do not reveal any evidence of life. The age of these rocks are called azoic because no fossils of living being were found from these rocks. Some scientists believe that some very microscopic beings had come into existence, but it is very difficult to prove their existence. Therefore, this age is also known as Archaeozoic age.
2. *Proterozoic Age*: In this age, we have definite evidences of origin of life. The time period of this age has been established as 120 crore years to 55 crore years from today. The creatures of this age were like Jelly fish. They did not possess bone, skin and hide. Their remains are not found from sedimentary rocks, but their existence can be assessed indirectly. Even today, there are such microscopic organisms whose evidences about their existence will not be available to future geologists. These early beings originated probably in shallow water. Plant kingdom also originated in this age as algae like plants. The earliest aquatic creatures and plants were ancestors of

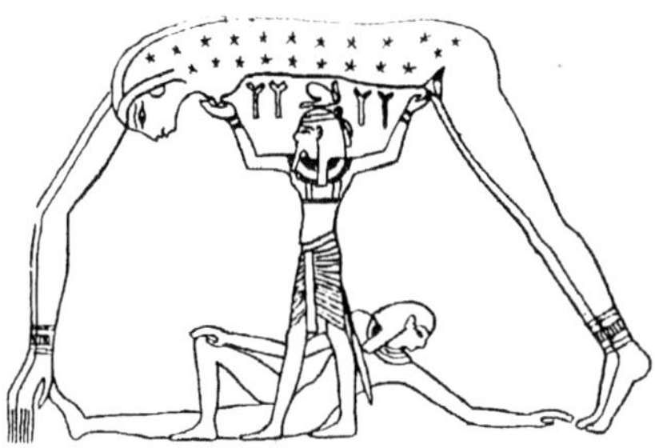

Fig. 2.1: An Imagination of Creation of Inhabitants of Prehistoric Egypt

"In the beginning God created the heaven and the earth. And the earth was without form and void; and darkness was upon the face of the deep. And the spirit of God moved upon the face of the waters."

all animals and plants belonging to animal kingdom and plant kingdom today. It can be taken as evidence that all living beings—plants and animals—need water for their survival even today in one form or the other.

3. *Palaeozoic Age*: The time period of this geological age has been established as 55 crore year to 20 crore year from today. It is also known as primary period. In this age, one finds the existence of such animals who possessed hide on their body to save themselves from the bitter sunrays. Such hideful creatures were small fishes, creeping insects, water scorpions and crabs. Water scorpion (Jal Bichhu) possessed a length of 9 feet. They were most populous creatures of Palaeozoic age. But after sometime, a change in environmental condition took place and one finds sudden more and more increase in number of fishes on the earth. In these fishes, eyes, teeth and vertebral column had evolved. These fishes were 2 to 20 feet in length. Their number had so increased in this age that this part of Palaeozoic age is known as **Age of fishes** (Matsya Kalpa). Till the Age of fishes, the life was restricted only to water. The land was still in azoic age. At the end of Age of fishes, a great change in the environment of earth took place. As a matter of this great environmental change, the land become livable for plants and creatures. First of all, vegetation came out on land situated nearby the source of water. The land situated nearby the source of water possessed some water too. Along with vegetations, some creatures such as water scorpion, crab, frog, dragonfly, creeping creatures or reptiles came-out and began to live there along with vegetations. The creatures moving towards land situated nearby the source of water were partly aquatic and partly terrestrial. They are called amphibians. They had started breathing in air. But being aquatic they had not developed efficiency to live far away from the source of water. Like frog of today, they had to enter into water for laying their eggs. Their off-spring used to lead their early life in water after being librated from eggs. The vegetation of this age also used to spread their

roots in water. In spite of this fact, there were numerous vegetations in this age. We dig out their remains today from coal mines in the form of coal. The later part of this Palaeozoic age is also known as Age of carbon (Barban Kalpa).

Fig. 2.2: Fishes Breathing in Air of Palaeozoic Age

4. *Mesozoic Age.* The time period of this age has been established as 20 crore years to 6 crore years from today. It is also known as *Secondary period.* In the early part of this age several changes took place in the environment of the earth. As a result of this a greater part of plant and animal kingdom of Palaeozoic age were destroyed. They had become extinct. But in this changed condition, new efficiency was evolved in plants and animals. Now one finds new vegetations and animals living on earth. Now such plants appeared which yielded fruits and seeds. Such creatures also appeared which could reproduce their off-springs far away from the source of water. Now new creatures did not need to go to source of water for laying their eggs and bringing up their children. Now their young ones started breathing in air after their birth. Creatures belonging to this age were reptiles such as snakes, crocodile, lizards, etc. These reptiles are found even today. But some reptiles such as Giganotosaurus, Plesiosaurus,

Table 2.1: Geological Time Scale and Origin of Various Forms of Life

Age	*Period*	*Epoch*	*B.C.*	*Important Forms of Life*
Cainozoic Age	Quaternary	Holocene	10 thousands	Off-springs of Homo sapien sapiens
		Pleistocene	4-3 millions	Australopithe cene Java man, Peking man, Heidelberg man Swans-combe man, Steinheim man, Fontechevade man, Neand-erthal man Homo-sapiens sapiens
	Tertiary	Pliocene	1.2 crore	Ancestors of man associated with Eoliths
		Miocene	2.5 crore	Anthropidapes, dryopeithecus, Sivapithecus, Ramapithecus, Proconsul
		Oligocene	3.5 crore	Early anthropoid ape, propliopithecus and other mammalian animals
		Eocene	5.5 crore	Abundance of modern mammals
		Palaeocene	6 crore	Early mammals, tarsiers, lemuroids, tarsioids
Mesozoic Age	Secondary	Cretaceans	12.5 crore	Insectivorous, early mammals and aves
		Jurassic	16 crore	Toothed birds and flying reptiles
		Triassic	19.5 crore	Pterodactyles
	Primary	Premium	22 crore	Expansion of amphibians and end of trilobites
		Carboniferous	30.5 crore	Primitive reptiles, vast forest and terrestrial spiders
Palaeo-zoic Age		Devonian	35.5 crore	Amphibians, fishes, expansion of forest
		Silurian	39.5 crore	Sea scorpions, early vegetations growing on land
		Ordovician	48 crore	Early fishes, ostracoderms
		Cambrian	55 crore	Unknown life on land, trilobites, molluscs, brachiopods
Protero-zoic Age		Precambrian	1 billion – 20 crore	Metazoa
Archaeo-zoic Age			2 billion – 70 crore	Protozoa

Tyranosaurus and Diplodocus have become extinct. They were 70 to 100 feet in length. They used to rule over the land. That is why this age is also known as Age of Reptiles.

Fig. 2.3: Diplodocus of Mesozoic Age

5. *Cainozoic Age*: The time period of this age has been established as 6 crore years to present day. In the early phase of this age, the earth witnessed the incidence of heavy glaciation. Due to the incidence of glaciation, reptiles became extinct because they had not developed efficiency to survive in very cold conditions. The animals of this age had to face a great struggle for the existence of life. They had to adapt themselves with changed environment. The natural condition of earth during this age also got changed. Mountains like Himalaya, Alps, Andige, Rockledge came into existence and islands and oceans were created. One can find these natural creation even today. After this change one finds plants and animals existing in grass land and forests. Now new animals were evolved. These were aves and mammals. Some creatures of this age were also existing in Mesozoic age. When there was plurality of reptiles, many animals could not succeed in the struggle of existence for life. They migrated towards cold regions. The nature covered their body with feathers to protect them from severe cold. These feathers helped them to fly in the sky. These animals called aves had to lay eggs on land or in the hole of trees or caves. They had to take care of eggs properly till the birth of young ones. For this purpose, male and female birds prepared nest. Female lay eggs. Both sit on eggs in nest covering eggs with their feathers. Some other animals of this age developed hairs. The nature provided them hair to protect them from the cold. They began to lay

birth of young ones in place of laying eggs. Such young ones had to depend upon their mothers for several months for their survival. They began to survive on breast feeding and care of mothers after birth. These animals were called mammals. Female mammals began to possess two mammary glands for this purpose.

Fig. 2.4: Mammoth and Heidelberg Man of Cainozoic Age

During Mesozoic age or Reptile age, the number of mammals were very few. They were confined only to cold regions. But in Cainozoic Age one finds sudden increase in their number and types. In this age, they began to rule the earth in the same manner as Reptiles in Mesozoic age and fishes in Palaeozoic age. Camel, horse, ziraffe, yak, elephant, zebra, deer, dog, monkey, man, etc. evolved in this age on the earth. The origin of these animals was very significant development on earth during this age. Animals found till now were self-dependent after birth. Parents left them after their birth to survive them independently. Young ones did not possess any feeling with parents and vice versa. But mammalian young ones were dependent upon their parents for a long time. This gave birth to emotional relation between them. This relation was expressed not only through non-verbal symbols, but also through verbal signs. The speaking power of these animals were more than other animals. They expressed their feelings through sound symbols by pronunciation. Due to this, off-springs began to take advantage of experiences of their parents. This gave rise to consciousness in them. On the other hand, the feeling to reciprocal relation gave birth to social life and social feeling. These feelings were not possible for reptiles.

The scholars have divided the animals belonging to class mammalian into sixteen orders. Among them most advanced order is primate. This order includes ape, anthropoid ape and man. Females of animals belonging to this order possess two mammary glands on their breast. They give birth to young one in one time. The number of children does not go more than 2 to 5 generally in the whole reproductive age.

It is believed that ape like animals had come into existence nearly 4 crore years ago. They gave rise to hominid animals. Early evidence of existence of hominid was more than 5-6 millions. These hominids were not true homo sapiens. But they resemble homo sapiens to some extent in shape and structure. It is, therefore, believed that homo sapiens evolved from hominid animals.

Fig. 2.5: A Bird of Holocene Period

Cainozoic age occupies significant place from the point of view of evolution of man. The Cainozoic age is divided into two periods namely *Tertiary* and *Quaternary*. For the purpose of study the scholars have divided these periods into several smaller periods. The **tertiary** period has been divided into five parts which are as follows:

1. *Paleocene period*. The time span of this period has been established as 6 crore years to 5.5 crore years from today. The environment of earth was more hotter than present day in this period. The number of mammalian animals went on increasing but evolution of man had not taken place. Earliest lumuroids, tarsio and insectivores appeared on earth.

2. *Eocene period.* The time span of this period has been established as 5.5 crores to 3.5 crores. Abundance of modern mammals appeared on earth in this period.
3. *Oligocene period.* The time span of this period has been established as 3.5 crore years to 2.5 crore years from present day. In this period smaller primates resembling man to some extent had come into existence.
4. *Miocene period.* The time span of this period has been established as 2.5 crore years to 12 millions from present day. Himalayas and Alps mountain came into existence during this period. Man did not appear in this age.
5. *Pliocene period.* The time span of this period has been established as 12 millions to 4 millions years. In this period, animals resembling man had appeared on the earth.

Quaternary

Quaternary period has been divided into two parts which are as follows:

Pleistocene period: The time span of this period has been established as 4-3 millions to 10,000 years from present day. In this period major changes took place in the climatic conditions of the earth. Due to these changes the northern regions of the earth faced heavy glaciations four times. The first ice-age came into existence nearly 600,000 years before present today. After this the climate of the earth again became favourable. This period is known as 1st interglacial period. The second glacial period came into existence nearly 400,000 years before present day and third glacial period came into existence nearly 200,000 years before present day. Similarly two interglacial periods also came into existence. The fourth and last glacial period came into existence nearly 100,000 years before present day. After that environment became favourable. Today we are living in fourth interglacial age or post glacial age.

In Europe, the study of glaciations has been done in the regions of Alps mountain. Penck and Bruckner made their study in 1909 and propounded the concept of Alpine glacial cycle. They provided the four glacial periods with names Gunh, Mindel, Riss and Wurm. They also discovered that the Second interglacial period was greater than the period of other two interglacial. They called this as the great interglacial. When glaciations occurred in northern regions, south regions witnessed pluviations. In these regions four pluviations and three inter-pluviation periods had taken place.

It was during the period of Pleistocene that Hominids gave rise to Homo sapiens gradually. Palaeolithic periods of archaeologists run parallel to Pleistocene period.

Holocene or Recent Period. The time span of this period has been established as 10000 years before present day. This period is still continuing and we are living in this period. Later stone age and metal age of human civilisation have also been placed in this period. A detailed description about man of Pleistocene and Holocene periods and their culture and civilisation has been presented in Chapters of the book to come.

3

Position of Man in Animal Kingdom

Man's Place in Animal Kingdom

Division	:	Metazoa
Phyllum	:	Vertebrate
Class	:	Mammalia
Sub class	:	Theria
Section	:	Eutheria
Order	:	Primate
Sub order	:	Anthropoidea
Series	:	Catarrhine (old World Monkeys)
Family	:	Homnidae
Genus	:	Homo
Species	:	Sapiens

Man is a member of animal kingdom. Rather he occupies top most position of animal kingdom. He is master of animal kingdom because his physical features are superior to other animals.

For the purpose of study animal kingdom is divided into two divisions—Protozoa and Metazoa. Man is placed under the division metazoa because his body consists of more than one cell. Like protozoans, he is not unicellular. Rather man is a multicellular animal. So, he has been kept under the division metazoa.

The division metazoa is divided into two phyla: invertebrate and vertebrate. In the body of man, there is presence of vertebral column. So, he has been placed under the phylum vertebrata. Invertebrates do not have vertebratal column.

The phylum vertebrata is divided into five classes: fish, Amphibian, reptile, bird and mammal.

Fish posses gills for breathing and fins for swimming. Absence of gills and fins in man excludes him from the class fish. The nature and structure of reproductive organ of amphibians are quite different from man. Therefore, man cannot be placed under the class amphibia.

Reptilians are cold blooded animals and man is a warm blooded animal. Therefore, man cannot be placed under the class reptillia.

Absence of feathers is sufficient to exclude man from the class Ares or Bird.

Thus, naturally man belongs to the class mammalia or mammals.

Chief Characteristics of Mammals:

- Mammals are air-breathing animals.
- They posses epidermal covering in the form of hairs.
- They are warm blooded vertebrates.
- They are viviparous, i.e. they do not lay egg.
- The foetus is nourished from the blood system of mother through placenta before birth
- After birth, baby is nourished by breast feeding, i.e. milk of mother.
- The skull has two condyles on two sides of foramen-magnum for articulation with atlas.
- The lower jaw articulates directly with skull without any articulation of quadrate bone.
- Two sets of teeth—deciduous and permanent—are seen.
- Teeth are different in shape, i.e. heterodont.
- Teeth are embedded in shocket, i.e. thecodons.
- Each vertebra is ossified from three primary centres— two for the arch and one for the body.
- Intervertebral discs are present in the vertebral column.
- Pectoral girdle is formed with the vestige of coracoids.
- Each long bone has two epiphysis and a shaft.
- Iliosacral junction is periacetabular
- An epiglottis is present. It is movable plate of cartilage which guards the opening of larynx.
- The heart is complete four-chambered organ. It is divided into two halves. Each half consists of a ventricle and an auricle.
- The body cavity is divided into two: one anterior thorax and a posterior abdomen by a diaphragm. The diaphragm is partly muscular and partly tendinous.

- Lungs are freely suspended within the cavity of thorax.
- The ureters lead from kidney to the urinary bladder.
- The cerebral hemispheres are united by a corpus callosum.
- Two halves of the cerebellum are joined by ponsvarolii.
- The RBCs are non-nucleated circular discs.
- No cloacae is present.

The class mammal is divided into two sub-classes: Prototheria and Theria. Man has not been placed under the sub-class Prototheria because Prototherians lay eggs, hatch them out and then suckle the young one at the breast. This stage represents a transitional stage between reptiles and mammals. Protherians are spiny ant-eater and duckbill.

Man has been placed under the sub-class Theria. The sub-class Theria is again divided into two sections: Methatheria and Eutheria.

In Metatherians the placenta is either undeveloped or functions for a few days. The kangaroo is example of Metatheria. Methatherians are also known as Marsupials or pouched animals because the mother possesses a pouch in the abdomen where young ones are transferred after being born in a foetal condition and are nourished with the help of nipples.

Man has been placed under the section Eutheria because he is a placental mammal.

The section Eutheria is divided into sixteen orders: Among them nine important ones are: Edentata, cetacea, sirenia, ungulata, carnivora, insectivora, chiroptera, rodentia and primate.

Edentata. The members of this order are either toothless or show presence of degenerated teeth which lack enamel. For example ant-eaters, sloths, armadillos, etc. They are no close relative of man.

Cetacea. The members of this order lack clavicle. Forelimbs are paddle like. The body is fish like and devoid of hairs. The example is whale, a sea dwelling mammal. They are not closely related to man.

Sirenia. They are aquatic mammals and their bodies are porpoise like. They are very far from man.

Ungulata. The members of this order are four-footed and hoofed animals. They are herbivorous. For example— cattle, deer, horse, rhinoceros, etc. Man is not closely related to them.

Carnivora. The members of this order usually live on flesh. They do not have clavicle. Their digits are provided with claws. Examples are cat, dog, tiger, lion etc. Man is not closely related to animals of this order.

Insectivora. Members of this order are insect eaters. They are nocturnal in their habit. For examples shrews, moles, etc. They are not close relative of man.

Cheiroptera. The members of this order show modification of forelimbs into a peculiar structure supporting the wing. For example bat. Man is not closely related to bat.

Rodentia. The members of this order are small animals. They have furry or spiny covering. They are provided with claws. Example are squirrels, rats, rabbits, etc. They are not closely related to man.

Primate. This is most developed order of the class mammals. Man is a member of this order. The chief characteristics of this order are as follows:

- Limbs are prehensile and adapted for arboreal life.
- Either thumb or great toe or both are opposable.
- Existence of flat nails on digits. This makes the grasping functions of hands and feet easier.
- Presence of well-developed clavicle.
- Orbits are completely surrounded by bony rims.
- Teeth are adapted for mixed food vegetarian and non-vegetarian.
- Three kinds of teeth are present at least one time in life.
- The teeth are differentiated into incisors, canines and molars to serves different functions of cutting, holding, grinding, etc.
- Mammae are two in number and pectoral in position.
- Nursing habits of mother determine the position of memory glands.
- The development of brain show wide variation.
- A calcarine fissure and a posterior lobe is always present in primate brain.
- The testes are descended into sacrotum. This is present in many other agile and active mammals also.

- Primates posses pendulous penis. This character is shown by other mammals also.
- Presence of well-developed caecum a sac-connected with the intestine of an animal.
- The stomach is simple.
- The femur has never a third trochanter.

The above mentioned characteristics are not exclusively for primates. The order primate consists of more than 500 species which are classified into a number of genera, family, sub-family, and finally included under three sub orders: Lemuroidea, Tarsioidea and Anthropoidea.

Let us first of all discuss Tupaiidae or tree shrews because it stands at the base of primate stem and form a very interesting group.

Prosimian Primates

Tupaiidae or Tree Shrew: For some scholars, tree shrews are members of the order insectivora, but for some other scholars they are members of order primate.

Tupaiidae comprises single living family Tupaiidae which is divided into two sub-families. These again include six genera, 32 species and 100 sub-species. They are distributed in India, Burma, China, Borneo, Sumatra, Philippine islands, Malaysia, etc.

Characteristics

- The tree shrew are small animals.
- They are squirrel like in appearance.
- They are arboreal in their habits.
- Some of them are diurnal, while some are nocturnal.
- They are mainly insectivorous, but they eat anything that is digestible.
- Their typical dental formula is 1 2/2: C 1/1: PM 3/3: M 3/3.
- Their digits are provided with sharp curved claws.
- The Manus and Pes are adapted for grasping.
- Both pollex and hallux are opposable to some extent.
- Their tail is used as balancing organ.
- Their tail is not prehensile.

- Their skull resembles with that of Lemur in many respects.
- Their orbits are directed more laterally than that of Lemurs.
- The rhinarrium is naked, glandular and moist.

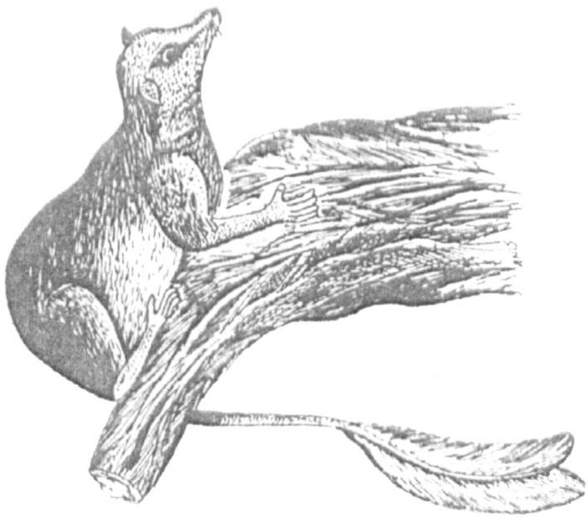

Fig. 3.1: Tree-shrew

Nature and Behaviour: Tree shrews are very active animal. They can make nearly 70 jumps in a minute. They spend their time mostly showing jumps and plays. They are also an aggressive animal. Males are more aggressive than females. They secrete smell symbols from glands of chest and breast. They rub their chest or breast glands with branches to secrete smell symbol. This behaviour is more clear just after fighting between two male individuals. Tongue and tooth combing have been observed. They are not found in big social group. As soon as male children attain younghood, they are killed by the adult male of the group. Females also kill their male children. Their pregnancy period is of 40 days.

Lemuroidea (Lemur)

This sub-order comprises five families, 17 genera and a large number of species. The families are: the Lories, the true Lemurs, the Indrises, the Galago, etc. The present day Lemurs are connected to old world. They are found in tropical forests of Africa and the islands of Madagascar. Some members are also found in Indonesian islands and Philippines (Fig. 3.2).

Fig. 3.2: Lemur

Characteristics of Lemur. The chief characteristics of Lemur are as follows:

- The Lemurs are small animals.
- Their size varies from that of a mouse to that of a large cat.
- They live on insects, fruits, leaves, shoots, nuts, egs, etc.
- They are arboreal.
- They are mostly nocturnal in their habits.
- They are coated with furry covering.
- They have long bushy tail.
- Tail is not prehensile, but helps in balancing.
- The hind limbs are longer than forelimbs.
- The hands and feet are adapted for grasping.
- All digits of hands and feet, except the second digit of hind limb, are provided with flat nails.
- Lemurs have naked rhinarrium and crescentic nostrils.
- The post orbital wall is absent.
- Eyes are directed outwards.
- The lachrymal foramen or tearduct is outside the orbit.
- A portion of palatine bone is present in the inner wall of orbit.
- The skull is not balanced upon vertebral column, rather it is suspended from occipital condyles.
- The external auditory meatus is absent.
- A large spherical auditory bulla is present.
- The jaw is elongated to give it fox-like appearance.

- The typical dental formula is 12/2: C 1/1: PM 3/3: M 3/3.
- In the lower jaw canines are incise form.
- The anterior premolars look like canines in the form.
- The canines and incisors are placed in front of jaw in a row like teeth of a hair comb.
- These are separate from first premolar by a gap.

Nature and Behaviour: Lemurs are tree-dwellers. They are expert in moving from branches to branches of trees. They are social by nature. Their sex ratio is 1:1. Tongue and teeth combing is present. The mother combs the hair of baby by her tongue and teeth. They move with the help of smell symbol. At the time of movement male and female Lemurs secrete smell symbol from their smell glands. This helps other members to follow the movement route. They recognise in-group and out-group members. On seeing out-group members, they make sound symbols to terrify the out-group members.

During morning period females are aggressive and males are submissive. Males are excited during evening and at night. Pregnancy period lasts between 120 to 135 days. Mating season comes once in a year. Baby Lemur is fully dependent upon mother. Baby remains adhered to the stomach of mother. After 10 weeks baby is able to walk around mother. Female Lemur plays a role of good caring mother.

Tarsioidea (Tarsiers)

This sub-order comprises only one living family Tarsioidea. It has a single genus Tarsius represented by a single species Tarsiers spectrum. They are confined to Malayan islands i.e. Sumatra, Java, Borneo, Celebes, Philippines, etc. (see Fig. 3.3)

Characteristics of Tarsiers Spectrum: The chief characteristics of Tarsiers spectrum are as follows:

- The Tarsiers are small animal.
- Their body attains a size of a rat.
- The body size of an adult is 8".
- They are usually insectivorous.
- They are completely arboreal.

Fig. 3.3: Tarsius Spectrum

- They are nocturnal in their activities.
- They can rotate their head for 180°. Thus, they can look backward without moving the body.
- Their movement is saltatory or frog like leaping movement.
- They have furry covering on body.
- Some parts of body are sparsely clothed with hair.
- The hind limbs are longer than forelimbs.
- The tail is longer and tufted at the tip.
- The second and third digits of the feet bear claws while other digits are provided with flat nails.
- The big toes and thumbs are opposable.
- The nostrils are complete and widely separated.
- The opening is marked by a narrow margin of naked skin.
- The post-orbital wall is present
- Orbits are very large and completely directed forward.
- The lachrymal foramen is outside the orbit.
- The bulla is large and inflated.
- The tympanic annulus is outside the auditory meatus.
- The typical dental formulae is 1 1/1: C 1/1: PM 3/3: M 3/3.
- The tibia and fibula are fused together at their lower ends.

Nature and Behaviour: They are found in couple. Mothers keep their babies adhered to stomach. They are good caring and nursing mother. They are tree dwellers. There is lack of sufficient information on their mating behaviour.

Difference between Tarsiers and Lemur: The Tarsiers differ from Lemurs in the following characters:

- In Tarsiers post-orbital wall is present, but in Lemur it is absent.
- In Tarsiers foramen magnum is more anteriorly situated than in Lemur.
- In Tarsiers occipital lobes are much more developed than Lemurs.

Difference between Lemuroidea and Anthropoidea: Lemuroidea is distinguished from Anthropoidea in the following characters:

- The post-orbital wall is absent in Lemuroidea, but it is present in Anthropoidea.
- Lachrymal foramen is outside the orbit in Lemuroidea, but it is inside the orbit in Anthropoidea.
- Eyes are directed outward in Lemuroidea, but they are directed forward in Anthropoidea.

Thus, many characters of man are not in a favour to put him under Lemuroidea or Tarsioidea. Naturally, man has been placed under sub-order Anthropoidea.

The sub-order Anthropoidea has been divided into two series: Platyrrhine (New World Monkeys) and Catarrhine (Old World Monkeys).

The series Platyrrhine (New World Monkeys) consists of two families: Hepalidae and Cebidae

Hepalidae: This family is represented by small squirrel sized marmosets. The members of this family possess the following characters:

- Their food consists of mainly fruits and insects.
- They are arboreal.
- They are diurnal in their habits.
- They are clothed with soft and silky fur.

- All digits except great toe are provided with claws.
- The great toes bear flat nails.
- The thumb is not opposable.
- The great toe is capable of some amount of mobility.
- The length of hind limbs exceeds that of forelimbs.
- They possess non-prehensile tail.
- The tail is tufted at its end.
- The nostrils are oval and widely separated.
- The muzzle is absent.
- The auditory bulla is large inflated chamber.
- The tympanic annulus is outside the cavity of bulla.
- The dental formula is 1 2/2: C 1/1: PM 3/3: M 3/3.
- The skull is small and round.

Nature and Behaviour: They live in small group. Male and female live in couple. They are tree dwellers. They give birth to more than one children. They have a pregnancy period of five months and one week. Children remain adhered to the stomach. After few months they ride on back. Children spend more time with father in comparison to mother. They spend time with mother for satisfaction of hunger.

Cebidae (Cebus): This family consists of a large number of genera and species which show diversified characters. They are found in the forest of South America and Central America. The members of Cebidae family differ from the member of Hepalidae family in following characters:

- They are comparatively larger in size than the marmosets.
- Most of them do not posses bushy tails.
- Their digits, except the first one, do not bear claws, but possess nails.
- The members of this family are Cebus monkey, Spider monkey, Squirrel monkey.
- They are diurnal.
- Their food consists of fruits, vegetables and insects.
- Their forelimbs are smaller than hind limbs.
- Their dental formulae is 1 2/2: C 1/1: PM 3/3: M 3/3.

Fig. 3.4: New World Monkey (Cebus)

Nature and Behaviour: They are found in a group of 4-5 members to 40 members. They recognise well-in-group and out-group members. They do symbolic war with outside members. They are cooperative with in-group members. Females deliver only one child at a time. Children are found adhered to their mother.

Only dental characters are sufficient to exclude man from the series Platyrrhine. Naturally, man belongs to the series Catarrhine (Old World Monkeys).

The series Catarrhine consists of three families namely, Cercopithecidae (Old World Monkey), Simidae (Anthropoid apes) and Hominidae (man).

Cercopithecidae (old world monkey): The chief characteristics of the family Cercopithecidae are as follows:

- They are arboreal in habit.
- They are herbivorous.
- They are also frugivorous.
- Only Baboons are terrestrial and to some extent insectivorous.
- The members possess tail of variable length—very short or very long.
- The tail is not prehensile.
- All digits are provided with flat nails.
- The ischial callosities are always present.

The Old World Monkeys comprise two sub-families: Cercopithecidae and Semnopithecinae

- The sub-family Cercopithecinae includes Celebes or Black apes, Macaques, Guenons, Baboons, etc.

- The Celebes apes are confined to Asia, China, Japan, Philippines, East Indies, North Africa, and Gibraltar. Baboons inhabit the region of South Africa, Arabia, Abyssinia and Guenons are found in different parts of Africa.
- The members of sub-family Cercopithecidae always have cheek pouches. All digits including the thumb are well-developed and provided with nails.
- The members of sub-family Semnopithecinae is represented by Langurs, Guerezas, retrousse-nosed Langurs, Proboscis monkeys etc. They are confined to Asia except Guerezas which are African variety.
- The members of sub-family Semnopithecinae have no cheek pouches. Ischial callosities are small. The thumb is reduced or absent. They have slender built bodies, short tales and long slender non-prehensile tails.
- The Semnopithecinae is distinguished from the Cercopithecidae by the presence of sacculated stomach and the absence of cheek pouches.
- The sacculated stomach is well-adapted to digest food consisting mostly of leaves.

Fig. 3.5: Old World Monkey (Macaque)

Difference between the Old World Monkeys and the New World Monkeys: The old world monkeys differ from the new world monkeys in the following characters:

External characters

- All species posses ischial callosities.
- All digits are provided with more perfectly flattened nails.
- The thumb and great toe are more highly specialised.
- They are capable of more independent and individual movement.
- The nostrils are situated close to each other.
- The tail is not prehensile.
- The Semnopithecinae have sacculated stomach and Cercopithecidae have cheek pouches.

Internal characters

- The bregma is placed relatively far forwards.
- The palate is relatively long.
- The basicranial axis is comparatively short.
- The auditory bulla is absent.
- The tympanic annulus is elongated and it forms an external auditory meatus.
- The dental formula is 1 2/2: C 1/1: PM 3/3: M 3/3.
- The upper premolars are three rooted and lower two rooted.

Anthropoid Apes: The anthropoid apes are represented by four main types: Gibbon, Orang-utan, Chimpanzee and Gorilla. They are tailless creatures and thus they are distinguished from most of the lower primates. They do not possess ischial callosities (excepting the Gibbon). In this character they differ from the old world monkeys. They are big-brained, flat-chested and long-armed animals.

This group consists of two sub-families namely Hylobatidale and Simidae or Pongidae.

The sub-family Hylobatidale consists of two genera Hylobates or the Gibbon proper and the Symphalangus. Similar consists of Orang-utan, Chimpanzee and Gorilla.

Hylobatidale (Gibbon): As referred to above, Gibbon family includes two genera: Hylobates or Gibbon proper and the

Symphalangus or Siamang. The Siamang is represented by only one species and is found only in the islands of Sumatra. But Gibbons comprising several species are distributed in various parts of eastern Asia including Assam and Burma and also in islands of Indo-Malayan Archipelago. By nature, they are tree dwellers, they do not make nests. But sometimes they move on land. They also move near the source of water to quench thirst. They also quench their thirst from drops of water present on leaves of trees where they generally live. The main physical characteristics of Gibbon are as follows:

- Gibbons are smallest of apes in size.
- In erect position, the length of Gibbon is 3 ft.
- Average weight of an adult common Gibbon is 13 pounds while that of Siamang is 24 pounds.
- They live in small family consisting of parents and children.
- They live on leaves, flowers, fruits and insects.
- Bird's eggs and young birds are also eaten by them.
- The Gibbon posses such long upper extremities that when it stands erect, the finger tips touch the ground.
- The extraordinary length of arms is directly associated with the brachiating mode or locomotion followed by Gibbons about 90 per cent of time.
- On the ground Gibbon may walk in the upright position using the arms for balancing.
- The skin colour is usually black.
- The whole body except the face and ischial callosities is covered with fine woolly hairs.
- The head is narrow and elongated.
- The digits except the thumb bear elongated and laterally compressed nails.
- The thumb is imperfectly opposable.
- The foot is long and narrow.
- The great toe is set far apart from the other digits.
- The great toe is capable of wide range of movement.
- The digital formula both in Manus and Pes is 3>2>4>5>1.
- The cranium is enlarged and somewhat flattened.
- The facial portion is reduced in size.
- The cranium presents ovoid contour and low frontal elevation.
- The orbits are large and prominent.

- The supra orbital torus is absent.
- The nasal bones are flat and broad.
- The nasal aperture is ovoid.
- The mastoid and styloid processes are not developed.
- The glenoid fossa is shallow.
- The post-glenoid tubercle is well-developed.
- The bregma is situated far back on the roof of the cranium.
- The dental characters show sexual differentiation.
- The canine is large, sharp and pointed and interlocks on closure of the jaw.
- Common Gibbons differ from the Siamang in some characters such as Siamang is larger than Gibbon, it has shorter legs and longer arms, trunk is relatively shorter, chest is broader, skull is longer, cranial capacity is large, webbing of the second and third toes is more advanced, and it has a laryngeal sac.
- The skull of Gibbon is distinguished from other Anthropoid apes by its absolute smaller size. It differs from that of lower Anthropoidea by the absence of two transverse ridges crossing the crowns of moler teeth. It may be differentiated from that of Cercopithecidae which has longest nasal bones and smaller skulls with lesser cranial capacity.

Fig. 3.6: Gibbon

Nature and Behaviour: Gibbons live in a small group. The group consists of an adult male, female and children. The number of children remains usually four. The children leave the groups after attaining adulthood. Sometimes, old male and female are found living as a couple. Tongue and tooth combing between male and female and between mother and children is common phenomenon. There is no fixed mating season. It may take place any time in a year. Sometimes, mating has also been observed during pregnancy period. The pregnancy period lasts for 210 days. Female delivers one child in one time. The birth space between two children is of two years. The baby is fully dependent on mother for nearly two years. The mother keeps baby in between folded thigh and stomach. The baby holds the waist of mother tightly.

Orang-utan (Pongo or Simia): Orang-utan is also known as Pongo or Simia and is represented by only one species. They are found in the islands of Sumatra and Borneo. They possess the following characteristics:

- They measure about four feet in height from vertex to sole.
- The males are heavier than females.
- The average weight of males is 165 pounds and that of female is 80 pounds.
- They are arboreal animals.
- Their movements are said to be very slow and deliberate.
- On the ground they go on all fours.
- They are very slow, sluggish and morose creatures.
- Their diet consists of wild fruits, leaves and shoots.
- They live in family consisting of a male, female and young ones.
- Long, coarse, shaggy hairs are sparsely distributed on their body.
- Their hair colour is reddish brown.
- Their face, ears, soles and palms are naked.
- Their skin colour is brownish.
- Nostrils are oval or slit like in-out line. They are placed obliquely.

- In profile, the nose gives a flat, concave appearance as root is narrow and bridge is not elevated.
- The face is also concave in profile.
- This condition of face is described as sinognathism.
- The ears are small.
- The hand is long and narrow.
- The fingers except the thumb are elongated. The thumb is much short.
- All digits bear flat nails.
- The digital formula is 3 > 4 > 2 > 5 > 1.
- The foot is also long and narrow.
- The great toe is very small and opposable.
- The upper extremities are very long.
- In erect posture finger tips reach the ankles.
- The skull is distinguished by its relatively small braincase and enormous facial portion.
- The average cranial capacity of male is 416 c.c.
- The supraorbital ridges are not continuous and no torus is formed.
- The forehead is high and rounded.
- The nasal bones are extremely small and fussed together even in young animals.
- The mastoid process is not developed.
- There is no styloid process.
- The glenoid fossa is shallow.
- Post-glenoid tubercle is large
- The external auditory meatus is not much prolonged.
- The jaw is enormous size and is projecting.
- The chin is absent.
- The dental characters show sexual differentiation.
- The canines are long, sharp, tusk like and interlocking.
- They have 12 pairs of ribs as in man, as against 13 pairs in Gibbon, Gorilla and Chimpanzee.
- The skull of Orang-utan is larger in size than Gibbons.
- The Orang-utan differs from Gorilla and Chimpanzee by large jaw in relation to the skull and splint like nasal bones, sinognathous face and non-continuous supera orbital ridges.

Fig. 3.7: Orang-utan

Nature and Behaviour: They live in small group. They are tree dwellers and cave dwellers. Their body weight is very much. They move on tree with great care. During childhood, they make jump on trees from branches to branches. On land, they follow quadrupedal walk. Sexual dimorphism is clearly visible in them. They prepare nests on trees. Adult males spend their times independently. Females are accompanied by children.

Chimpanzee (Anthropopithecus): Chimpanzee are found in the tropical forests of Africa. They include three species—common Chimpanzee, bald-headed Chimpanzee and pygmy Chimpanzee. Their chief characteristics are as follows:

- The average height of males is five feet.
- The average height of females is four feet.
- The average weight of males is 110 pounds.
- The average weight of females is 88 pounds.
- They live in small bands of about 8 or 9 individuals.
- Their food consists of various kinds of fruits and vegetables.
- They also eat ants.
- They use grass stick or wood stick in the hole of ants for eating them.
- They are expert in climbing and brachiating.

- They run about on all four.
- They are capable of walking and standing erect.
- Their body is somewhat sparsely covered with long, coarse, shaggy hairs of variable colours.
- Their skin colour also varies.
- In typical Chimpanzee the margins of nostrils are not much expanded.
- Ears are very large but without lobes.
- The hand is elongated and narrow.
- The fingers are long.
- The thumb is opposable.
- The palm is naked.
- The foot is also long and narrow.
- The opposable great toe is set apart from other toes.
- The digital formula is $3 > 4 > 2 > 5 > 1$ both in Manus and Pes.
- They are short-legged and long-armed animals.
- Finger tips go down the knee in standing position.
- The vault of head is low.
- The cranial capacity of male is 400 cc.
- The supra orbital ridges are continuous and prominent, but not so massive as in Gorilla.
- The nasal bones are much reduced in length and are fussed in the early part of life.
- The nasal bridge is not elevated.
- The external auditory meatus is prolonged and tubular.
- The glenoid fossa is shallow in comparison to that of Gorilla.
- The post-glenoid region is well-marked.
- The jaw projects forward.
- The chin is absent.
- The canines are large, sharp and projecting but in lesser degree in comparison to Gorilla and Orang-utan.
- The Chimpanzee skull may be distinguished from Gibbon by its absolutely large size.
- In Chimpanzee, supra orbital ridges are continuous but they are discontinuous in Orang-utan.
- In Chimpanzee, nasal bones are not splint like and the lower jaw is relatively small, but in Orang-utan nasal bones are splint like and the lower jaw is big.

- Chimpanzee differs from Gorilla in having smaller skull with relatively larger cranium, absence of prominent crowns and numerous other details.

Fig. 3.8: Chimpanzee

Nature and Behaviour: They live in small groups. They prepare sleeping place each day with branches and leaves. They do not sleep at the same place on next day. They are wandering animals. They wander in search of food during day time. They prepare sleeping place in the evening and spend night over there. In the morning they again start wandering. They select sleeping place at a height of 15 ft or more. They are vegetarian, but they eat ant with great taste. They prepare tool of grass or wood stick for this purpose. They also train their children in preparing tools for ant eating. In a group of Chimpanzee, one finds three to seven members. They are very clever animal. They also perform art. They are used in circus for showing plays. Males are more submissive. Mating of seven males from the same female was observed. Infighting among males is not observed during mating with the same female. When two Chimpanzees meet, they wish each other by touching head, shoulder, thigh and sex organ. Teeth and tongue combing has also been observed. They attain adulthood in seven to eight years. Menstrual cycle continues after 37 day. Pregnancy period is of 227 days.

Gorilla

The Gorilla are inhabitants of western and eastern central equatorial Africa. They comprise one genera and two sub-species namely,

Western Coastal or Low Land Gorilla and eastern or mountain Gorilla. The Gow Land Gorilla are found in the Cameroon and the Gabon. The Mountain Gorilla inhabit in the region of eastern Congo, west of Lake Edward and Lake Kiou. The chief physical features of Gorilla are as follows:

- The Gorilla has an enormously robust body.
- The average weight of body is 450 pounds.
- The height ranges from 5 ft to 6 ft in Low Land Gorilla and 5′3″ to 5′10″ in mountain Gorilla.
- The Gorilla pass most of their time on ground.
- They very often climb on tree also
- Upon the ground their usual movement is quadrupedal though they can stand and walk erect.
- They walk upon the sole of foot.
- They are slow moving animals.
- Their diet consists of fruit, shoots, leaves and vegetables.
- They live in small bands of three to ten families.
- The skin colour is black to dark brown.
- The whole body except face, palm and sole is covered with black, long, coarse hairs.
- The ears are small in size and bear no lobe.
- The nose shows slight elevation at the bridge but the root is low and flat.
- The nasal wings are very thick.
- Nostrils are broad and directing forward.
- The lips are thin.
- The hands of Gorilla resembles with man in certain characteristics such as thumb is better developed and hand is comparatively shorter and broader than other anthropoid apes.
- The foot of Gorilla is adapted for arboreal prehension and it shows a tendency to resemble human foot.
- The great toe is opposable and set apart from other digits.
- They are short-legged and long-armed creatures.
- The finger tips reach below the knee when Gorilla stands erect.
- The proportion of arm to forearms of Gorilla resembles with man.
- The head is very large.

- The cranium is smaller.
- The face and jaw is enormous and projecting.
- The supra orbital ridges are of enormous size and continuous to form a torus.
- The forehead is very low.
- The skull presents a very prominent digital crest which is weaker in female than in male.
- The females are also smaller in body size.
- The dental characters also show sexual differentiation.
- The average cranial capacity of male is 550 cc and of female is 450 cc.
- The mastoid process is moderately developed.
- The jaw is large.
- The chin is absent.
- Canines are of enormous size.
- The teeth are large, more highly and more sharply cusped than those of Chimpanzee.
- There are also differences between Mountain Gorilla and Low Land Gorilla. The Mountain Gorilla have a narrower skull, a large face, longer palate, shorter limbs, shorter and broader hands, large trunk, narrow hips etc.

Fig. 3.9: Gorilla

Nature and Behaviour: Gorilla perform their activities mostly on land. Sometimes females and children climb upon trees. They prepare nests on ground from branches and leaves. They reside in their ground nests only for a night. Next day they prepare another nests for their stay. They prepare sleeping nests within five minutes. They are vegetarian by nature. They do not use any tool. Low land Gorilla eat fruits. The High Land Gorilla eat forest climbers, barks, leaves, bamboos, etc.

The Low Land Gorilla also use banana and sugarcane. They live in a small or a big group. Their group may consists of five to thirty members. Usually, there are two females for a male. Each group has an adult male leader. Reciprocal teeth and tongue combing is observed. Grooming is also observed between mother and children. Children and adolescents play together. After attaining an age of six years, they start maintaining a distance.

They do not have fixed mating season. Meating can take place in any season in a year. Menstrual cycle begins after 30-31 days. The pregnancy period lasts between 251-289 days. They deliver only one child in one pregnancy. Babies are dependent upon mother. Breast feeding continues between 6 to 9 months. The maximum breast feeding period was observed for one and half year. Babies are adhered to mothers for three months. After three months, they begin to walk upon the back of mothers. After six months children begin to walk with mothers. Mothers spend four days with a dead baby. Only then they leave it.

Hominidae

In absence of pre-maxilla bone, presence of chin and in many other details a man differs from Cercopithecidae and Anthropoid apes. That is why man has not been placed under the family of Cercopithecidae and Simidae (Anthropoid apes). Therefore, man has been placed under a separate family Hominidae with one genus Homo and a single species Sapiens (Fig. 3.10).

Some of many distinguishing characters of man or true diagnostic characters of Homo as a type are as follows:

General Characters

- Bipedal walk.
- Fully erect posture.
- Large size and complex character of brain.
- Greatest weight at birth in relation to body weight in adult life.

Fig. 3.10: Modern Man

External Characters

- The density of hair (except on scalp) is reduced. Thus, the body is relatively hairless.
- Tactile hairs are completely absent.
- The margin of ears are highly rolled.
- The nose is prominent.
- Elevated bridge is formed by the nasal bones.
- The fleshy tip is supported by osteo-cartilaginous framework.
- The median furrow or philtrum is present on upper lip.
- The mucus membrane of upper lip is visible as the latter is outrolled.
- Ischial collosities are absent.
- Peculiar relation of skin to the subcutaneous fat which is adherent to the formers.

- The legs are longer than arms.
- The foot is arched both transversely and anteroposteriorly.
- The great toe is not opposable, it is in the same line with the other toes and is under the common skin covering approximately up to the mid point of the basal phalanx.

Skeletal Character

- The cranium is enlarged and the face is reduced.
- Shortest height of face.
- Nasal bones are large and broad.
- Nasal bones are separated by intro-nasal suture.
- The pre-maxilla is never marked off from the maxilla by a suture.
- The spheno-maxillary fissure is wide.
- The external auditory meatus is relatively short.
- In the orbit the lachrymal articulates with ethmoid bone.
- The sphenoid and the parietal bones are articulated at the side of the wall of brain box.
- The ethmoid articulates with sphenoid in articular fossa.
- The mastoid and styloid processes are well-developed.
- The foramen magnum is placed far anteriorly at the base of the skull so that the head is well-balanced on the spinal cord.
- The chin is present.
- The simian mandibular shelf is absent.
- No special sexual differentiation in the teeth is seen.
- The upper molars are single rooted or double rooted.
- The lower lateral incisors are broader than the central and the upper centrals are broader than the laterals.
- The canines are small. Their rows are always at the same level with those of other teeth.
- Diastema is absent in the dental series.
- Highest average total number of vertebra.
- The highest average number of thoraco lumber and coccygeal vertebrae except Gibbons.
- The lumbar curve of the vertebral column is fully formed.
- Pelvis bone is absent.
- Linea aspera in the femur is well-developed.

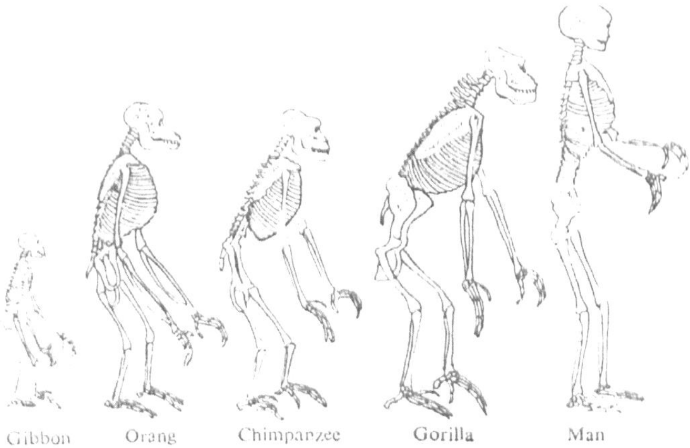

Fig. 3.11: Vertebral Column of Monkeys, Apes & Man

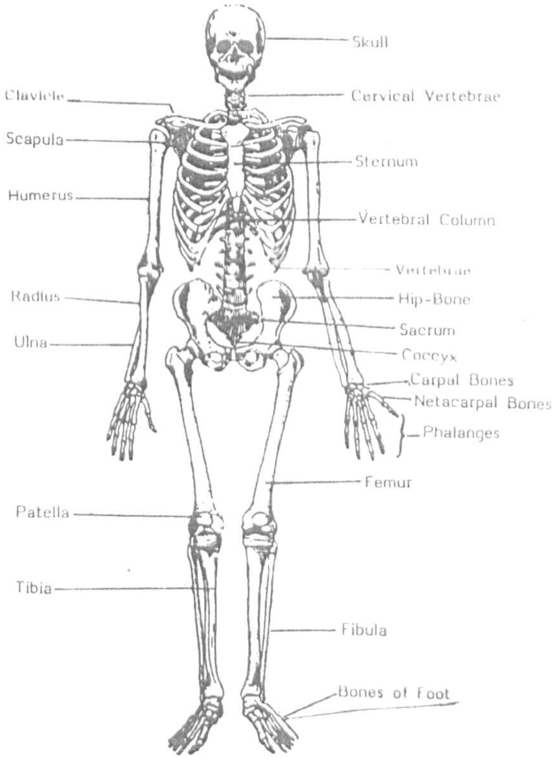

Fig. 3.12: Human Skeleton

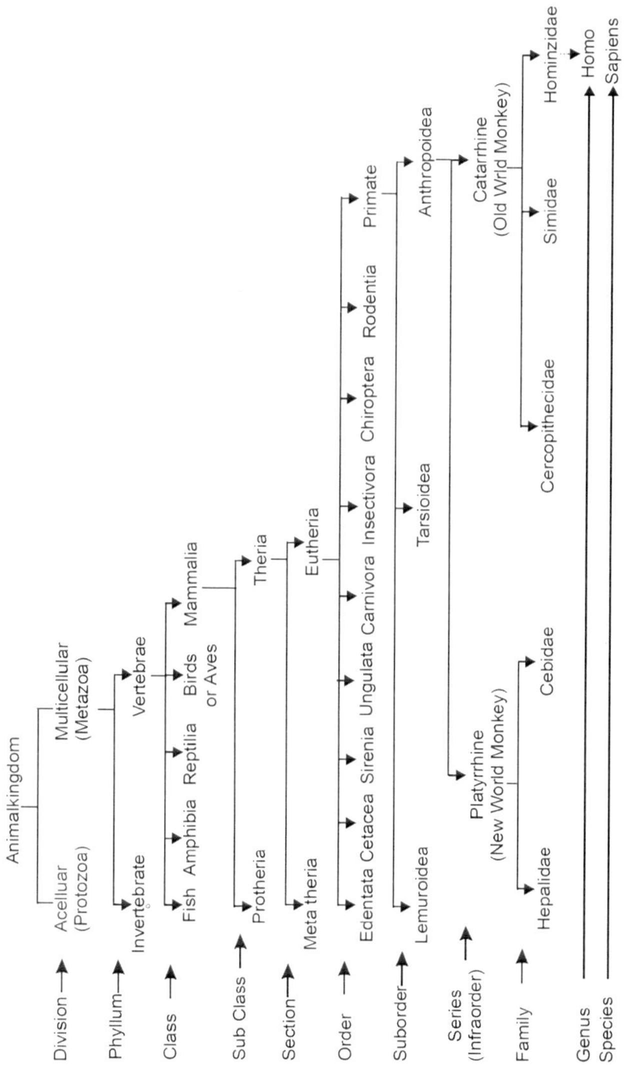

Chart 3.1: Showing Position of Man in Animal Kingdom

4

Theories of Biological Evolution

Biological evolution is nothing but development of dissimilarities between the ancestral and the descendent population. The dissimilarities are created by changes in characteristics through evolutionary processes. The emergence of genetics has brought about many changes in the concepts of human evolution. Now human evolution is looked from the point of view of genetics. Genetically human evolution may be defined as changes occurring through frequency of genes in human gene pool. But in the study of human evolution, morphological characters cannot be ignored. Evolution is also necessarily be viewed as changes in morphology of organism through time. Morphological characters include many variables like size, shape and colour of the skin.

According to evolutionist anthropologists man has emerged from some non-human primate stock. Man had to pass through various developmental stages to attain his present position. His present position distinguishes him from other organisms. In order to understand the various stages of evolution, the early varieties hominid have been studied and their phylogenetic positions have been established.

The present concept of man's ancestry is based on studies of such long-term evolutionary adaptations. The study of evolutionary process deals with factors responsible for evolutionary changes.

Now Let us Deal with Various Theories of Biological Evolution

Various theories propounded by different scientists and scholars to explain the evolution of organism on the earth are as follows:

- Lamarckism
- Neo-Lamarckism
- Darwinism
- Neo-Darwinism
- Synthetic Theory
- Macro-Evolution and Micro-Evolution

Lamarckism

Jean Baptiste de Lamarck (1744-1829) was a great French philosopher and biologist. He suggested a complete theory of evolution. He propounded his theory of organic evolution in his book philosophic zoologique (1809). His theory is known as Lamarckism. His theory is based on the following points:

1. *Spontaneous generation and increase in size*: According to Lamarck, the forces of life tend to increase the size of individuals or its part up to a limit determined by life itself. This inherent tendency is found in all living beings.
2. *Direct effect of environment*: Environmental factors like temperature, water, land, air and food have a direct effect on the modification of structure of an organism. Due to change in environment new needs are created. New needs result in change of habit. Change of habit brings modification in the structure of organism.
3. *Use and disuse of parts*: According to Lamarck, in changed environment new needs result in modifications of some structures. These are used more and more than other organs. The more and more used organs are acquired and less and les used organs are reduced. They become extinct ultimately. This can be understood by the following examples:

   (i) The ancestors of modern Giraffe did not possess strong and enlarged neck. They were short necked animals. They used to graze on land. Scarcity of grass forced them to survive on leaves of trees. For this purpose, they started elongating their necks to catch the leaves. This habit resulted in more and more use of neck. As a result, the neck became strong, enlarged and functional.

This changed structure was acquired by the descendants of the Giraffe. Present day Giraffe has long neck.

(ii) Ancestors of amphibians were aquatic fishes. Some of them started carwling on land. They needed arms to crawls and lungs to respire in the changed environment. This gave rise to fishes with lungs and lobed fins. Now there was no use of gills, so they disappeared slowly. Later limbs and proper lungs developed as a result amphibians came into existence.

(iii) Due to constant use, the Biceps of hand of Goldsmith is developed.

(iv) In man right hand is more developed due to constant use.

(v) Due to disuse, the limbs of snakes have disappeared. Its vestige can be found in Pythons.

(vi) Due to constant use, the pina of rabbit is increased in size.

(vii) Due to constant disuse, the teeth of birds have disappeared. They have only groves.

(viii) Presence of teeth in whale embryo suggests that due to constant disuse, the teeth have vanished in whales.

All these examples reveal that organs which are in constant use develop properly and give rise to acquired characters.

4. Formation of new organ takes place as a result of need. Lamarck believed that new organs arise or old organs get modified. The new or modified organs are best suited to the environment.
5. Inheritance of acquired characters. According to Lamarck the acquired characters, i.e. new or modified structures are transferred from ancestors to descendants. That is why, descendants are better suited to the environment.
6. Formation of new species. According to Lamarck inheritance of acquired characters gives rise to new species.

The summary of Lamarckism can be enumerated as follows:

(i) Increase in size of organism and its component parts up to a certain limit is caused by inner forces.

(ii) Change of environment has direct effect on organisms.
(iii) This results in new habits.
(iv) New habit leads to use and disuse of organs.
(v) Organs which are in constant use develop properly and give rise to acquired characters.
(vi) Organs which are in constant disuse become extinct.
(vii) Acquired characters are inherited.
(viii) Inheritance of acquired characters results in formation of new species.

Criticism of Lamarckism: Lamarckism met with great objections from the very beginning of its publication. The main objections are as follows:

1. The statement of first law of Lamarckism is correct to some extent because lower organisms are capable of spontaneous generation and there is tendency to increase in size in some forms with evolution. For example, horse, camel and elephant. There are also some organisms which show decreasing trend due to evolution.
2. Lamarck second law of use and disuse of organ is not acceptable. For example muscles of a weight lifter becomes very strong and developed, but muscles do not develop if weight lifter leaves exercise. This acquired character is not transferred to the off-springs.
3. The statement of Lamarck on the formation of new organ due to effect of environment and new requirement is baseless. In animals, this is influenced by nervous system. For example, the lengthening of neck of Giraffe is due to its increase in size.
4. His statement that acquired characters are inherited, seems incorrect. For this purpose a number of experiments were performed. Experiments gave either inconclusive or negative results. Some of experiments are as follows:

   (i) *Kammerer's experiment* on Salamanders reveals that acquired characters are inherited. Salamander has yellow patches on black skin. He kept salamander in yellow boxes. He found that yellow patches increased in area.

When these were mated, off-springs with large yellow patches were produced. But when other scientists conducted the same experiment, they did not find the same result. The scientists found some injecting material in the laboratory of Kammerer. He ultimately committed suicide. However, it was found that due to change in environment, there was no change in the genetic constitution of animals.

(ii) *McDougal's experiment on rats.* McDougal on his experiment on rats showed that acquired characters were inherited. He trained a group of rats to escape through dark hole instead through a lighted whole with electricity. He observed that those trained rats showed increasing trend of learning from one generation to the next generation. In this way he found that acquired characters were inherited.

(iii) *Agar's experiment.* Agar and his associates performed experiment on albino rats. They trained one set of rats and used other set of rats as controlled group. They observed slight difference in the number of right choice of holes up to 15 to 16 generation. They disapproved McDougal's experiment in support of inheritance of acquired characters.

(iv) *Mutilation.* Weismann gave a hard blow to the theory of inheritance of acquired characters. On the basis of experiments he showed that cutting off the tails of newly born rats for successive 80 generations, did not bring about change in tails of succeeding generations.

(v) *Feet of Chinese woman.* Small feet of women in China is considered a beauty. The feet of Chinese women were kept bounded for generations, but they failed to produce small feet. This shows that acquired character is not inherited.

(vi) *Ear and nostril boring.* The custom of ear boring and nostril boring is being practised in Hindus since centuries. But these acquired characters have not been inherited. This is also against the theory of inheritance of acquired characters.

(vii) *Doctor's son*. The son of a doctor is not born a doctor. Similarly, sons of an engineer, IAS, IPS and Professor are not born as engineer, IAS, IPS and Professor.

(viii) *Wrestler's son*. The son of wrestler is not born with exceptionally developed muscles.

From the above account, it is clear that Lamarckism has to face severe criticism. The main objections labelled against Lamarckism are as follows:

(i) Lamarckism is not applicable to lower organisms in which cellular differentiation has not taken place. It is also not applicable to higher forms.

(ii) Morphologists have not accepted the views of Lamarck.

(iii) Weismann was correct in saying that new organisms develop from germ cells of parents which are effected by somatic cells or natural environment.

(iv) The body of an organism has both germ and somatic cells. New individual develops from germ cells which are not effected by somatic cells.

(v) Lamarckism has no experimental proof.

Neo-Lamarckism

Faith in Lamarckism was revived by the scientists like Herbert Spencer, Karl Marx, Engels, T.D. Lysenko, Tower, Agar, Geoffroy St. Hilaire, and Edward Steels. These scientists expressed their views in Lamarckism and attempted to revive it. Therefore, they are known as neo-Lamarckissts and their theory is known as Neo-Lamarckism. Thus, modified version of Lamarck's theory is called as Neo-Lamarckism.

A large number of experiments were performed which make us believe that a body character is result of interaction of heritable genes and environmental conditions. Environment does not influence the operation of genes. With the result new and varied characters appear in off-springs.

Generally, Neo-Lamarckists were of opinion that environment has direct effect on somatic cells which influence the germ cells. An interesting example was cited for this. The pituitary gland along with other gland also influences the gonads and so the germ cells. So, there is no doubt that other body cells are also capable of influencing germ cells. From our present knowledge of genetics we cannot accept that. It is accepted that pituitary influences the physiology of gonad, but not the genes present in germ cells. When genes are not changed, modifications acquired by somatic cells cannot be inherited.

Tower kept the immature potato-beetles in optimum temperature and moisture on somatic cells. This resulted in the modifications in their structures which were inherited in next generations. This was also not due to direct effect of temperature and moisture on somatic cells but due to changes in chromosomes.

Agar could produce a group of aquatic (simnocephalus) insects in a culture of flagellated protozoans with many modifications. When these were next developed in normal water, the off-springs also had the same changes. But these modifications disappeared from generations to generations. Agar concluded that in adults the body and gonads are affected due to abnormal conditions.

Likewise when X-ray is given to individuals, it has been found that it affects the chromosomes of germ cells and somatic cells of the off-springs. These experiments support Lamarckism. But even today, a concrete experiment is not known to us which can prove Lamarckism without a flaw.

Edward Steels stated that somatic mutations are beneficial and give rise to modification of part in which these cells are found. He also believed that DNA or genes of these cells are transported into nuclei of germ cells and are incorporated in the genetic materials. If it is true, then characters may pass from somatic to germ cells. But there is no evidence to support this view.

Darwinism

Charles Darwin (1809-1882) was a naturalist. He grew up in England in Victorian era. As a young man he wanted to study medicine, but he soon learned that he could not stand by the side of the blood. So, he dropped out of medical school in Edinburgh and enrolled in Christ's college, Cambridge. At Cambridge he became a budding naturalist and was encouraged by a number of faculty mentors. One mentor helped arrange his appointment as naturalist on the ship HMS Beagle when set out on a five year voyage around the world in 1831. The voyage of the Beagle was a crucial for Darwin's ideas.

Before the Beagle left for England, Darwin had begun reading Lyell's book, Principles of Geology (1830-1833). During the voyage, he completed the reading this work. He became inspired to search for a biological process equivalent to uniformitarian process of geology. As the Beagle sailed around the atlantic and pacific coasts of South America, Darwin observed that geographical distributions of varieties of plants and animals correlated with the distribution of variation in useful environmental resources. When he visited the Galapogs islands off the coast of Ecuador, he observed that varieties of finches and tortoises differed slightly from island to another and also differed from varieties on South American main land. How and why did these differences develop?

When the Beagle arrived back in England, Darwin was already converted to the idea of evolution. He undertook years of scientific research to strengthen his reputation as naturalist, while he pondered new, non-Lamarckian mechanism that might make evolution work.

Then, Darwin read Thomas Robert Malthus Book, 'An Essay on Principle of Population' (1798). Malthus was the pessimistic political theorists who explained how the human population of the world was increasing geometrically (2,4,6,8,10 etc.), while global resources needed for human survival were increasing only arithmetically (2,4,8,16,32 etc.). The inevitable consequence of these trend was that not everyone born could possible survive Darwin embraced this Malthusian vision and broadened it to include all biological nature. In biological nature organisms engage in struggle for existence producing survival of the fittest.

Knowing, now, how evolution works, Darwin began to draft his book on evolution. He worked on the book off and own for many years until he received a letter from his fellow naturalist Alfred Russel Wallace, (1823-1913). While writing from Pacific South Seas, Wallace described a theory of evolution by natural selection. Darwin recognised it immediately as almost exactly like his own. After consulting with his friends, he decided to finish his book quickly. First, Darwin and Wallace presented a joint paper on evolution to a meeting of Linnaeum society in London in 1858. The following year in 1859, Darwin's origin of species appeared. Ever since, the theory of evolution by natural selection independently formulated by both Darwin and Wallace has been known as Darwinism.

A good way to understand Darwin's theory of evolution is to contrast it with Lamarck's theory using the example of long-necked giraffe. In contrast to Lamarck, Darwin approached this example with a different premise. A group or population of ancestral short-necked giraffes was living on Savana like grassland. They needed to eat vegetation from trees to survive. Some giraffes had slightly longer necks than others. These giraffes had a slight advantage over other giraffes in the competitive struggle for vegetation needed to survive. Beating out the competition because of this natural advantage, they ate more, became healthier or in some other way had more off-springs. Gradually, overtime, long necked giraffes had more off-springs than the short-necked giraffes. Average neck length in the population increased, until the present long-necked giraffe spices evolved.

Darwin presented this sequence of events as natural selection. Its meaning metaphorically that nature selects advantageous traits just like human breeders artificially select advantageous traits when they domesticate plants and animals. The result in both cases is that organisms become adapted to their environment except for the result of plant and animal breeding, almost all of these evidence was circumstantial. He argued that anatomical and embryological similarities among organism, the presence of vestigial organs, and the record of fossils, though incomplete were all consistent with the theory.

A problem for Darwinism, then and now, is that this same evidence is consistent with many versions of creationism. Eventually, the scientific community came to accept Darwin's theory. His theory represents an extension of scientific revolution from astronomy and physics into biology. Darwin really went beyond Newton, because he showed that basic structures of the universe evolve.

Origin of species provoked a barrage of moral, religious and social criticism. Many critics failed to realise or admit that the book made hardly any reference to the evolution of Homo sapiens. Darwin ducked this controversial topics for several years. Some of his friend confronted the controversy head on. The main implications for Homo sapiens was the evolution of human mental and moral qualities.

Most Christians believed that animals lacked spirituality and were mentally and morally a world apart from human beings. Could evolution bridge this gap. In the geological evidence of the antiquity of man (1863) Charles Lyell described human evolution as natural leap on to a new plan of life. Alfred Russell Wallace disagreed arguing that mental and moral superiority would have conferred no real selective advantage on animals. Therefore, could not have evolved in the first place. Why would an animal need to be artistic, mathematical or philosophical. According to Wallace, divine intervention must have been responsible. Other scientists were more open to the idea of Darwinian human evolution. At the same time only a few human fossils were known. Unfortunately for human evolutionists, these fossils appeared neither particularly old nor particularly primitive.

Thomas Henary Huxley in his book 'Evidence as to Man's Place in Nature' (1863) classified people and apes in the same taxonomic order. He was staunch supporter of Darwin. Therefore, he was nicknamed as Darwin's Bulldog. He had defended Darwin staunchly in public debates. Without fossils, the artifact record of pre-history became more important. So, human evolutionists also cited the work of archaeologists like John Lubbock and cultural evolutionists like Lewis Henry Morgan and Edward Burnett Tyler.

Eventually Darwin published his views on human evolution in the Descent of Man (1871). Much of this book and also of Darwin's the 'Expression of Emotions in Man and Animals' (1872), was devoted to the argument that differences between animals and people are differences in degree rather than in kind. To explain the evolution of human physical traits, Darwin used the mechanism of sexual selection. Sexual selection traits evolve not because they confer an additive advantage in the struggle for existence. But because, they make members of one sex more attractive to the other and in this way increase reproductive success.

According to Darwin, Human intelligence was evolved by natural selection, as by-product of upright stature. This freed human hands for the use of the tools. To explain the evolution of human morality. Darwin relied on mechanism of Group Selection. According to Darwin, the core of morality was altruism. Altruism is willingness to sacrifice oneself for the good of others. Altruism was initially selected in groups, when one member behave altruistically and as a result, other group members were benefited. Later when human beings became intelligent, they extended altruism beyond the local group to all humanity in the form of abstract moral order.

Viewed from the perspectives of modern science, Darwin's explanations appear to conflate or confuse biological and cultural evolution. In this regard Darwin was not much different from his Victorian contemporaries. Almost all Nineteenth century evolutionists were extremely hereditarian. Like racism in archaeology, racism in biological anthropology was a legacy from the Nineteenth century.

Darwin and his friends did not expose many of the religious, moral and social attitudes now labelled as Darwinian. The main religious challenges to Darwinism was not based on Biblical fundamentalism because by 1960s, Bible was no longer widely accepted as necessarily historically accurate. The main religious challenge was based on morality. If human beings were product of evolution, not divine creation, would not a system of morality have to be based on the process of evolution itself. And if so, would not the easiest way to construct such a system be to treat evolution as intrinsically and ultimately purposeful. Contrasted with Lamarckian evolution Darwinian evolution appeared to lack ultimate purpose. It operated opportunistically selecting characteristics adapted to only circumscribed time and place. Alternatively, if Darwinian evolution were a divine instrument—Gods way of creating, the mechanism of natural selection appeared excessively brutal. It involved relentlessly hard struggle, competition and death of individuals unable to adapt. It was always possible, of course, to argue that these unfortunate losses were compensated for by evolutionary winners who helped humanity improve. But this position was morally precarious. In most cases it was easier to abandon Darwinian evolution in favour of Lamarckian mechanism of inheritance of acquired characteristics. It was most humane and offered hope that people might take charge of their evolutionary fate. In the late Nineteenth century, Lamarckism became the doctrine of choice for majority of scientists seeking to reconcile evolution with religious morality.

In the discussion of social morality, the term social Darwinian is historically misleading. Most of the social attitudes denoted by this term derive not from Darwin, but from Herbert Spencer, the most philosophical and sociological of the classical cultural evolutionists. It was Spencer who had used social Darwinian most popularly. He used it most often to rationalise social inequalities among races, classes and genders. Meanwhile, Huxley, Darwin's Bulldog advocated an opposing version. Huxley was agnostic who actively doubted religion and believed that science should maintain moral neutrality. He opposed Spencer and anyone else who based social morality on biological evolution. According to Huxley Homosapiens has evolved through cosmic accident at a point where people are able to understand that evolution has no purpose. Why not take advantage of this opportunity and create a morality that is independent of evolution and even goes against harshness of nature. In the Nineteenth century, between extremes of Huxley and Spencer, there were so many versions of social Darwinism that the term really needs to be defined almost every time, it was used.

Amidst all the wrangling over religious and social morality, Darwin's theory of evolution by natural selection suffered major scientific setbacks. From beginning, there had been much experimental proof that natural selection could produce new species even with artificial breeding, which produce many sub-species or varieties. Another problem was the fossil record. Darwin admitted that the record was imperfect and contained gaps or missing link. Some scientists attempted to fill these gaps with speculative evolutionary sequences such as those based on biogenetic law. This law stated that ontogeny, the growth of an individual, recapitulated phylogeny, the evolutionary growth of species. Proponents of the law made extreme statement about embryological and palaeontological similarities and detracted from the credibility of evolution as empirical science. Yet another problem was the age of earth. Evolution by natural selection was a slow process that required a great deal of time to account for changes observable in fossil record. Contemporary physicists, thinking about volcanic activity as an agent of geological change, decided that the earth had been much hotter in the past than in the present and that volcanic activity had been much more forceful. A troubling implication was that this volcanic activity had wrought geological changes too quickly for Darwinian evolution to have worked.

A final problem for Darwin was swamping effect, the name given to the observation that small variations serving to raw material for natural selection would always be swamped out through heredity, preventing natural selection from ever getting started. Darwin was aware of these scientific problems as a result he grew discouraged. He lost confidence in the complete efficacy of natural selection. He was so discouraged that in later edition of origin of species, he turned to other evolutionary mechanism, including the Lamarckian mechanism of inheritance of acquired characteristics.

The solutions to Darwin's scientific problems was beyond the scope to explore in technical detail. What Darwin needed was the theory of biological heredity pioneered, unknown to him, by Austrian monk George Mendel (1882-1884), the father of modern genetics. Principles of Mendelian genetics were worked out in 1860s, discovered by scientists in 1900 and incorporated into the synthetic theory of evolution in 1930s. Until then, in biological circles the theory of evolution by natural selection remained theoretically weak. In anthropology, however, the shadow of Darwinism loomed strong as it conjured up much what twentieth century theoreticians sought to reject.

Darwin's Theory of Natural Selection

Darwin's theory of natural selection is based on the following facts:

1. *Over Production of Off-springs*: All living organism sproduce much larger number of off-springs than can possibly survive. A rabbit produces six young ones at a time, four times a year. All young ones become sexually mature after six months. An oyster produces 6,00,000 eggs per year. Ascaris produces more than 15,000 eggs per day. If all off-springs a single species survive and reproduce without check, the particular species will cover the entire world. But the number of individuals of a species roughly remains constant. This is because of the reasons given in next coming points.
2. *Limited Supply of Food and Shelter*: Though the off-springs are produced in large numbers, yet food and shelter available for them remains always constant.
3. *Struggle for Existence*: On account of the above fact, a fierce and active competition is taking place for the survival. Every

individual (organism) tries its best to get the maximum food and shelter for itself. The struggle for existence may be—

(a) *Intra-specific.* Struggle between members of the same species to gain supremacy.
(b) *Inter-specific.* Struggle between different species living at a common place.
(c) *Environmental.* Struggle for existence with environmental conditions like draught, flood, extreme heat and cold, disease, earthquakes, etc.

4. *Survival of the Fittest*: In the struggle for existence only those organisms survive who are best suited to the environment. The weaker organisms becomes extinct. As a result, the number of organisms on the earth remains always constant.
5. *Variations*: Due to constant struggle and changing environment organisms change themselves. Due to this no two individuals are exactly alike. According to Darwin, variations are one of factors of organic evolution.
6. *Inheritance*: Those characters which are useful for the individuals to survive are transmitted to the off-springs through reproduction in next generation. That is why off-springs are more suited. For this, he gave the theory of Pangenesis.
7. *Natural Selection and Origin of New Species*: On the basis of above facts, Darwin came to the conclusion that in course of long periods, the organisms make themselves more and more adjusted to the nature. And due to inheritance of variations in the successive generations, new species are formed. In the beginning, these newly formed species do not show much difference from parent species. But in course of time and changing environmental conditions diverse characteristics appear. In this way, from a single ancestor more than one new species may arise.

Criticism of Darwinism

Although Darwinism can explain to some extent the mechanism of evolution, but most of the scientists do not agree with the view of Darwin. There are some views in its favour, and some views are against it.

The facts supporting the Darwinism are as follows:

1. By artificial selection, it is possible to produce various domesticated varieties.
2. The number of ancient animals decreased due to change in environment. The animals of the present day have some favourable characteristics which helped them to survive in changed environment and unfavourable surroundings.
3. The evolution of horse provides strong support of Darwinism. The modern horses evolved as a result of modifications of its legs and teeth due to change in environment. Evolution of cow, elephant, deer, etc. also suggest strong proof about Darwinism.
4. When black, white and checkered coloured chickens were allowed to move freely in a field, it was found that hawks spared the checkered coloured chickens as their colour matched with the background. This experiment was performed to support Darwinism by Davenport.

Facts disapproving the Darwinism are as follows:

1. Darwin gave importance to slow and small variations but could not explain the usefulness of the organs in incipient stage (organ not yet perfectly developed).
2. He was silent over use and disuse of organs and the presence of vestigial organs.
3. Darwin explained the survival of the fittest but not the arrival of the fittest.
4. He did not differentiate between the somatal and germinal variations and said both were heritable, which is not true.
5. How some species when crossed produce sterile off-springs whereas when varieties are crossed fertile off-springs are produced.
6. His theory could not explain the origin of coordinated structures to produce the huge antlers of the Irish Elk (deer). These antlers are hindrance rather than a help.
7. Many differences seen in the species are not at the adaptative

value, but are the result of incidental effect of genes. How and why should these be inherited, Darwin could not explain.

8. In order to explain origin and transmission of variations, Darwin brought forward his theory of Pangenesis. According to this view, small particles called Pangenes which through blood reach egg or sperm from modified parts are responsible for modified changes in the next generations. But this theory is not convincing and we know that characters are passed on through hereditary materials to the next generations.
9. Many Darwin's examples for variations are found today to be not heritable. Only those variations are heritable which occur due to change in genes.
10. Whether instincts are acquired and modified through natural selection, he did not say anything.
11. He could not speak about the evolution of Homosapiens.
12. He could not solve the problem of morality associated with evolution.

Most of the modern biologists do not accept the views of Darwin. Darwinism does not explain the presence of vestigial organs as well as mutations that arise during the lifetime of an individual. Further, the fittest characteristics are not always inherited by individuals of the next generations. For example, man with a very good body does not produce a child with good body. Natural selection cannot explain the universal occurrence of variations but accept this as a factor of evolution. Natural selection is not only factor for origin of a new species because mutation can also produce a new species.

Neo-Darwinism

In the light of recent results obtained from recent researches, Darwin's theory has been modified. The modified version of Darwin's theory is termed as Neo-Darwinism. The scientists who have modified the Darwin's theory are called as Neo-Darwinists and their theory is known as Neo-Darwinism. They are Weisman, Mendel, Devries, Fisher and J.B.S. Haldane. The theory of Neo-Darwinism is based on the following facts:

1. Variations caused by rearrangement of genes during sexual reproduction.
2. Variations arise by mutation.
3. Natural selection.
4. Genetic drift.
5. Isolation.

Struggle for existence was an important factor for natural selection. The most important objection of this view is variation which Darwin look as heritable. These variations may be due to environment or due to change in heritable factors. Mendel showed that characters depend upon gene action present on chromosomes. Genes are heritable units and always remain pure. But animals in which sexual reproduction takes place show different kinds of variation due to random assortment and recombination of genes. Mendel showed that mutations take place in genes which produce new characters. Therefore, characters for natural selection leading of organic evolution are produced by mutation and recombination of genes.

George Johan Mendel formulated the science and basic law of genetics in 1865. He obtained many varieties of peas and grew them in his garden. He scientifically obtained results of his cross-breeding. He published his conclusion in 1866.

Mendel's work remain unknown for 34 years. Significance of his work was recognised only in 1930, when Law of Segregation was discovered by three different biologists. Deveries in Holland, Correns in Germany and Tschermark in Australia. Since then great advances have been made in our knowledge of heredity based on original discoveries of Mendel. Breeding of plants and animals has become purely a scientific method these days.

Fisher was of opinion that due to natural selection beneficial genes have a tendency to dominate over less beneficial allelomorphs. Mutation has a definite rate which is called mutation pressure. The intensity of natural selection is known as selection pressure. Mutation pressure, selection pressure and genetic drift all along with other factors bring about evolutionary changes.

According to R.A. Fisher and J.B.S. Haldane, the mechanism of selection is based on particular inheritance.

Synthetic Theory

The twentieth century theoretical synthesis of Darwinian evolutionism and Mendelian genetics is known as synthetic theory of evolution. This theory was propounded by the Neo-darwinists. This theory is based on gene mutation, gene recombination, natural selection, genetic drift and isolation. Besides these five factors, other factors influencing evolution are migration, hybridisation and chance. The basic postulates of synthetic theory are as follows:

1. The process of evolution is result of mutation, recombination, natural selection, genetic drift and isolation.
2. Mutation does not affect evolution process directly. It is also not immediate source of variation. It is an important source of a variation which fulfils gene pool.
3. For natural selection mutation is an important factor which gives rise to new species. It affects very less to the phenotype and provides base to polygenic or multi-factorial inheritance.
4. Natural selection plays significant role in micro-evolution. Its role in genetic variation is creative and progressive.
5. Reproductive isolation affects gene flow. In new environment, new species are formed.
6. Genetic drift is also responsible for evolution.
7. Micro-evolutionary factors through continuous process in geographical limitations create new species.

Criticism of Synthetic Theory

The main criticism labelled against synthetic theory are as follows:

1. This theory is helpful in explaining previous evolutionary theory but it fails in providing the clue of future evolutionary changes.
2. This theory treats convergence and hybridisation as secondary factors in evolution. But both these factors play valuable role in human evolution.
3. Migration cannot be treated as a different factors because migration can be explained by convergence and hybridisation.

4. This theory is unable to explain the adaptation of new variations.

Macro-Evolution and Micro-Evolution

Organic evolution is a continuous dynamic process. Biological anthropologists attempted to study the macro-evolution and micro-evolution. Macro and micro are two approaches on the basis of which evolutionary processes are explained. Macro-evolution is concerned with long-range history of evolution. But micro-evolution is concerned with short-range history, between two generations i.e. between father generation and off-spring generation. The main points of differences between micro and macro-evolutions are as follows:

1. The time span of macro-evolution is very long but the time span of micro-evolution is very short.
2. Macro-evolution is concerned with total change of species. Such types of changes are visible in animal kingdom in long-time span. Micro-evolution is concerned with changes in an endogamous group.
3. Macro-evolution can be shown through the study of fossils but micro-evolution can be shown by studying the living population.
4. Macro-evolution is concerned with developed and isolated groups of species, but micro-evolution is concerned with continuous changes that have taken place or that is taking place in a population.
5. Missing link is helpful in macro-evolution, but it is of no use in micro-evolution because it is concerned only with living human population.
6. In macro-evolution, the entire pre-historic human populations and their descendants are studied. But in micro-evolution small endogamous human populations and their off-springs are studied.
7. Macro-evolution follows unilinear sequence but micro-evolution follows multi-linear sequence.
8. Broken teeth, bones and skulls are evidences for reconstructing the evolutionary sequence of macro-evolution. But micro-evolutionary sequences are reconstructed on the

basis of haemoglobin, gene pool, blood groups, enzymes, metabolic actions and pedigree.

9. Macro-evolution is concerned with human palaeontology, whereas micro-evolution is concerned with population genetics.
10. Macro-evolution is based on past-human events, but micro-evolution is based on present human events. This gives us an account of changes taking place in genes which can be observed in an endogamous group. Natural selection, mutation, migration, mate selection and adaptation with environment play significant role in micro-evolution.

Explanation of Some Terms

While dealing with the principles of evolution some technical terms have been used. It seems necessary here to explain those terms one by one to make them understandable to the scholars.

Natural Selection

Natural selection is most important mechanism by which evolution takes place. This fact was established by Charles Darwin in his book 'Origin of Species' (1859). Natural selection may said to be some total of environmental forces which decide the chances for survival of a species. Within a species there are certain variants which are better suited to survive under various environmental conditions. That is why they leave more number of off-spring. The less well-adapted variants fail to do so. That is why they are eliminated by nature. In this way, according to the concept of natural selection, the better adapted variant increases in number and gradually replaces the earlier type. Thus, natural selection acts as mechanism of evolution.

Darwin noticed that there is a tendency in all organisms to increase in geometric ratio. The elephant is very slow breeding animal. But if a pair of elephant gives birth to six young ones in a life time of 100 years and if this rate continues, then about 19 million elephants will born out of the first pair in about 750 years. But this does not happen in practice. The number of given species remains more or less constant. Many individuals do not survive in each species because of various factors. The individuals have to struggle for existence.

Those who are better equipped are selected by nature and that is why they survive. This doctrine of Darwin is known as 'survival of the fittest'.

Now-a-days, some scientists prefer to speak the phrase 'Darwinian fitness', 'selective value' or 'adaptive value'. This fitness is measured by the reproductive capacity of a population of the organism. Therefore, we can say that the continuation of one lineage of an organism and elimination of another by certain process or agent is the product of natural selection.

When we talk of better filled or les adapted individuals, naturally a question arises as to how new variations arise in an organism. Darwin did not know the answer. But on the basis of advancement of knowledge in genetics, it was known that a mutation was responsible for producing variation. Thus, mutation provides the raw materials for evolution. This is the primary process of creating genetic variabilities. It does got accomplish evolutionary change. We are to think the naturally occurring forces that operate on the variations to bring changes leading to evolution. In this way, complex interacting factors are involved in the operation of natural selection.

Mutation

Mutation is the change in the genetic materials. The change produced through mutation is passed on from cell to cell. Genes are usually very stable. They make faithful copies of themselves for generation after generation. But sometimes in the structure of gene a change takes place which is inheritable. As a result, a gene gives rise to a new form of the old character. This is called Mutation. There are a number of different types of Mutation. But the two broad categories are: Genetic Mutation and Chromosomic Mutation.

In Genetic mutation, the change may occur in the structure or chemistry of gene. Genetic mutation may be due to loss of one or more gene. It is called deficiency. Again, there may be duplication of same gene. These two changes are in regard to the number of genes. There may be also mutation of gene due to change in the arrangement of genes by way of translocation and inversion.

Chromosomic mutation takes place due to changes in the number of chromosomes. For example, one entire chromosome of a homologous set may be missing. This phenomenon is known as monosomic. When there is more than two homologous chromosomes, it is called polysomic. Again, each chromosome of entire set may be represented singly. This is known as Haploid. When it represented by more than two homologous chromosomes, it is known as polyploids. Mutations occur with regularity rarely. Substances known as mutagenes increase the rate of mutation. Mutagenic substances include certain chemicals, x-rays and cosmic radiation. Fluctuation of heat may also produce mutation.

It is generally believed that most of changes produced by mutations are harmful from the point of survival. As an organism is outcome of a chain of natural selection to adapt itself. Therefore, the disadvantageous genes tend to disappear. The carriers may either fail to reproduce or die out. Nevertheless, mutation at a slow rate is self-selectively advantageous. The possibility of changes by mutation helps the evolutionary line to continue adjusting itself to the changing environment.

Genetic Drift

We have discussed about natural selection and mutation as factors responsible for changing gene frequency. In the absence of these two factors, there is an another factor which brings changes in gene frequency. This is known as genetic drift. Genetic drift is also known as sampling error because it is a random statistical error and not a biological process.

Sometimes Genetic drift is also known as Sewell Wright effect after the name of population geneticist Sewell Wright, who described the process.

The process of Genetic drift is more effective in very small populations. In large population this process cannot play effective role. Therefore, small size of population is a pre-condition for the process of genetic drift to become effective. Further, it becomes more effective in an isolated population.

Let us consider a case of small population, where there are 100 individuals. Of 100 individuals, 95 are of blood group O and reset 5 are of blood group B. Because of factors like accident, illness, non-availability of mating partners, etc. the individuals belonging to blood group B may be gradually eliminated. In this process, increase or decrease of a particular allele does not depend upon advantageous or disadvantageous condition. This takes place merely as an accident or changes. Therefore, sometimes it is referred to as chance fluctuation or accidental loss or fixation of gene.

Let us consider another example. Suppose there was a small population of brown-eyed individuals. A few members of which were blue-eyed. Those blue-eyed individuals migrated to some other locality and settled there in isolation to give rise to another small population of blue-eyed. From parental population the blue-eye was lost accidentally. But in new population that gene was fixed.

Migration

Migration is also one of factors responsible for changes in gene frequencies in a population. As a result of migration genes are introduced to a breeding population from outside. In this way migration is responsible for admixture. Migration as such does not produce any new trait but the genotypes evolved in a certain geographical area find their way in another population elsewhere through migration. Thus, genes infiltrate from one population to another through migration. The intra-group genetic variability within a species tends to increase and inter-group genetic variability tends to decrease as a result of migration.

In America, coloured Africans and while Europeans migrated. Admixture took place between these two populations. This admixture gave rise to a mixed population. While Europeans migrated to South Africa where they came in contact with the local Hottentot population. Due to admixture of these two kinds of population, a mixed population emerged.

Isolation

Evolutionary changes are produced due to isolation. Isolation may be geographic and reproductive. A species may live in various environments. They may undergo different adaptations to those varied environments. As a result of different adaptations in different environments, several sub-species or races are created. Such a species is called Polytypic species. In course of time one sub-species may become distinctive to form a separate species due to isolation.

Parallelism and Convergence

Sometimes two organisms show close similarities between them in certain morphological patterns. The similarities may be due to parallelism, convergence or close phylogenetic relationship.

The term parallelism is usually applied when animals of same order or common origin show development of similar adaptive feature in course of evolution, though they belong to separate species. For example, the present day crocodiles and some fossil type of crocodile of reptilian age descended from some primitive reptiles. Both had to pass through distinct evolutionary stages. In course of evolution, they underwent similar changes. The absence of tail in man, great apes and Gibbon may also be due to parallelism. All of them had common ancestors, who had tails. The tail was lost in course of evolution in separate lines but in parallel fashion.

The term convergence is applied when two species develop similar characteristics, though they are not related by common ancestry. For example, the humming bird and the humming moth show convergence in their flying habits. They follow a common way of searching nectar in flowers. Nectar is their food. The wings of bats and birds show convergence. Stream lined shape of whales and fish shows convergence.

In some cases it is very difficult to suggest whether the similarities are due to convergence or parallelism. Organisms develop certain characteristics to survive under the same environment.

Homology and Analogy

While dealing with parallelism and convergence, it seems necessary here to discuss the homology and the analogy. It is because homology is related to parallelism and analogy is related to convergence. It is because of this fact that parallelism is also known as homologous evolution and convergence is known as analogous evolution. The wing of bat and the forelimb of monkeys are homologous because both show common ancestry. But the wing of bat and the wing of butterfly are analogous because they do not show common ancestry; though they perform the same function. Thus, analogy is similarity in function not in origin or ancestry. Homology is opposite to analogy. Homology is similarity in origin or ancestry, not in function.

Adaptive Radiation

Adaptive radiation is spread of a species by way of evolving progressively in to distinct type of organism. Thus, from a single ancestral form different new forms may be evolved to adapt to the rapidly changing environmental conditions.

Let us consider some examples of adaptive radiation. We know that reptiles are cold blooded animals and mammals are warm blooded animals. When climate became colder during the Pleistocene period, dinosaurs could not adapt to the changed environment. Therefore, they became extinct. But mammals evolved into distinct types to adapt to the new environment. Among mammals, primates began to live on tree, bats began to fly and survive in air, whales became adapted to survive in sea environment, carnivores became specialised in hunting, the rodents became adapted to grazing. Each of mammalian orders again developed into different forms, each form acquiring a new mode of life to adjust itself to the changed environment. In this way, mammals radiated into different parts of the world.

It is fitness of the thing to mention here that morphological characters of fossils and comparative study of living forms help us to understand adaptive radiation.

Irreversibility

A Belgium Palaeontologists of 19th century Luis Dollo was the first to suggest in 1893 that evolution is irreversible. That is why irreversibility is also known as Dollo's principles. In course of evolution, a change in structure of forms takes place. That changed structure is never reverted to its original form. Thus, in evolution a reversion to the original ancestral form does not occur. This does not mean that re-evolution of certain character similar to an ancestral form never occurs. For example, certain extinct flying reptiles possessed wings. In birds and bats wings were re-evolved independently.

Extinction

In the process of evolution, disappearance of certain species takes place. This is termed as extinction. A species may be wiped out totally or it may give rise to some later species. Thus, extinction is caused in two different ways. The change in environment may be a factor of extinction. A particular species may be well adapted to a certain environment. If due to some reason that environment is changed to such an extent that particular species find it difficult to survive. In this situation, the species is bound to become extinct. This happened in case of woolly mammoths. Wooly mammoths were adapted to glacial environment. But when the climate became warmer the wooly mammoths found it difficult to survive. They could not adapt themselves with changed environment. They became extinct ultimately.

In the second case, the species may be transformed to give rise to a new form of species at a later period. Here we can consider the case of Australopithecines, which is an extinct form. But this does not mean that the gene pools of Australopithecines have disappeared completely. Some genetic materials might have contributed to the development of Homosapiens.

5

Pleistocene Period and Its Main Features

The long period of time, through which the world has passed, has been divided into a number of major divisions or era of life. These are Archaeozoic (Archaean), Petrozic (Early), Palaeozoic (Primary), Mesozoic (Secondary) and Cenozoic (Tertiary and quaternary). Each eras has a number of epochs.

The tertiary period has gone further divisions into Palaeocene, Eocene, Oligocene, Miocene and Pliocene. Each division has a number of sub-divisions with which we are not concerned at the moment, though these are important in the consideration of various pre-human stages of man; the evidences of which have been found in these deposits.

The quaternary which is the last of the eras of life on the earth has two epochs, namely Pleistocene and Holocene. As man has originated in Pleistocene, so it can be said that pre-history also began in the same period. The enviroment of Pleistocene is closely related to cultural equipment and way of life of man. As such, it is not easy to understand any human culture without understanding the environment. Man's bio-cultural evolution cannot be understood without the knowledge of his climatic condition. The climate of Pleistocene began about four million years back from present time and continued up to 8,300 BC when Holocene epoch began and is continuing till today.

Pleistocene is known as the age of man. It is a term of Greek origin. 'Pleisto' means most and 'Koinos' means new. This term was used for the first time by Charles Lyell (1797-1875), a geologist in 1839. The beginning of this epoch is characterised by drastic changes in the environment. The homogenous warm climatic conditions characteristics of pre-Pleistocene epoch were no more seen in Pleistocene epoch. Pleistocene epoch was marked by Arctic ice region, temperate Europe and warmth Africa. In this period, the climatic change from warmth to cold and *vice-versa* were noticed. Due to this climatic variations, it is known as Great Ice-age.

Before dealing with the main features of Pleistocene, it will be fitness of the thing to know Pliocene-Pleistocene boundary and Pleistocene-Holocene boundary.

Pliocene—Pleistocene Boundary

Till recently it was thought that the Pleistocene epoch was one million years old and that the first glaciation occurred immediately after the epoch began. Regarding this there was no general agreement and thus this led to the confusion. In this process, deposits of similar age had been described by some workers as those of upper Pliocene, while some others have assigned them to lower Pleistocene. It was in 1912 that Haug defined Pleistocene as that epoch which yielded the evidence of mammals of one or more of genera, Elephas, Equus or Bos. This suggestion was supported and recommended for adoption by International Geological Congress held in London in 1948. A faunal assemblage of this kind is generally identified as Villafranchian. Villafranchain is recognised as a period of time which began about 3.5 million years ago. The term Villafranchian was used originally in Europe for the rock laid down on the land (rather in sea) after that time and before the large scale continental glaciations. As a consequence of this, change in the limit of Pleistocene, the interpretations that put man in upper Pleistocene had to be revised and placed in lower Pleistocene period. It is now generally accepted that the total duration of the Pleistocene divisions is more than three times of what was put forth earlier. The Villafranchian stage of the Pleistocene according to the present thinking includes pre-Gunz (Donau) glaciation.

Pleistocene-Holocene Boundary

There is no common agreement among the scholars regarding when Pleistocene came to an end and when Holocene began. Some scholars are of the view that Pleistocene epoch ended with the final retreat of the glaciations and with comparable disappearance of pluvial in tropical areas. But workers who consider fossil fauna opined that the advent of domestication should be taken as mark of end of Pleistocene. According to this school of thought the beginning of Holocene is put to 5000 BC, whereas according to earlier school of thought, Holocene terminates around 10000 BC.

Today majority of scholars accept the event of climatic changes to define the end of Pleistocene.

Pleistocene Period and Its Stratigraphy

On the basis of stratigraphical arrangement of rocks in the Alps mountains four main advances of glaciations have been recognised by two scientists, whose names were A. Penck and E. Bruckner. The four main advances of glaciations are called Ice-ages. They have been named as Gunz, Mindel, Riss and Wurm. These are widely adopted as Ice-ages of the Pleistocene. These four Ice-ages were intervened by three periods of normal conditions. They were called as Gunz-Mindel Interglacial period, Mindel-Riss Interglacial Period and Riss-Wurm Interglacial Period. The glacial periods indicated cold climate, while Interglacial period showed warm climatic condition. This does not simply mean that glacial and interglacial periods were characterised by cold and warm climates respectively. Some scientists believed that even during glacial periods some warm climatic conditions used to prevail. This view is supported by a warm climate animal called Hippopotamus which lived in Europe. According to G.G. Simpson the Antarctic climate in one of Ice-ages was 4°C more than that of present day. During the same Ice-ages in different parts of the world, the climate was warmer. But majority of scientists believed that temperature in all seasons, particularly in summer was low that is why the Ice-age with intense cold begins.

During the Ice-ages, the ice on the mountain tops located in polar regions and temperate zone spread very rapidly. In the North pole regions it used to be as thick as 3000 metres. In North and South America ice sheets occupied almost half of the continent. In North Pole regions it used to be as much thick as 3000 meters. In North and South America, the ice sheets occupied almost half of the continent.

In Europe, ice precipitated on hill tops of Scandinavia had spread up to the cities of London, Berlin and Warsaw. Even in Asia, minor evidence has been found to show that the ice sheets spread to Mount Olympus in Armenia and to Mount Lebanon and Mount Hermal of Palestine and Syria. But the climate in these areas seemed to be warmer as shown by fossil remains of such animal species as Hyrax, Cervas dama, Panthera pardaus, Crocuta, etc. These are evidences to show that the mountains of Pamer, Hindukush, Karakoram and Himalayas were also covered by ice-sheets.

In Africa, the ice spread was far less than in other areas. Here the tops of mountain like Mount Kenya and Kilimanjaro were covered by ice-sheet.

There is no clear cut and convincing evidence in support of the four Ice-ages for determining the age of Pleistocene epoch in various parts of the world. Even in Northern Europe, no evidence of Gunz was found. However, the geologist C. Gagel in 1913 on the basis of weathering horizons as evidence of Inter-glacials and on their fossil evidence in Northern Europe (North and West Germany), Britain and Scandinavia), Denmark, Poland and Russia, established three major glacials and two Inter-glacials as follows:

Weichsel Glacial	=	Wurm
Eemian Interglacial	=	Riss-Wurm Interglacial
Warthe, Salee } Glacials	=	Riss Glacial
Holstein Interglacial	=	Mindel-Riss Interglacial
Elster Glacial	=	Mindel Glacial

In the beginning it was believed that Pleistocene began during Gunz glacial period. The pre-Gunz Villagranchian strata were believed to have belonged to Donau period. But on the basis of pollen grain evidences from Southern Alps, it was felt that the climate was cold. Hence Donau period can be counted as the beginning of Pleistocene. The beginning of Pleistocene in Donau period brought four Interglacial and five glacial in the Pleistocene epoch. The period between Donau and Gunz were warm, so it was an Interglacial period as shown by the fossil evidence of an ape species called Macaca Florintina found in Holland.

Till recently when the Villafranchian was considered as the stage of Pliocene, the Pleistocene was divided into lower Pleistocene, middle Pleistocene, and upper Pleistocene. But with the inclusion of a Villafranchian stage in Pleistocene, now Pleistocene includes four stages namely, basal Pleistocene, lower Pleistocene, middle Pleistocene and upper Pleistocene. The beginning dates of these stages of Pleistocene are as in Table 5.1.

Table 5.1: Pleistocene Stratigraphy and Geological Time Scale

<table>
<tr><th>Epoch</th><th>Time approximately in years before present</th><th>Glaciation</th><th>Fauna</th><th>Division of Pleistocene</th></tr>
<tr><td rowspan="5">Pleistocene</td><td>100,000</td><td>Wurm Glaciation
Riss-Wurm Interglacial</td><td>Eemian</td><td>Upper Pleistocene</td></tr>
<tr><td>200,000</td><td>Riss Glaciation
Mindel-Riss Interglaciation</td><td>Hoxnian</td><td>Middle Pleistocene</td></tr>
<tr><td>400,000</td><td>Mindel Glaciation
Gunz-Mindel Interglaciation</td><td>Cromerian</td><td>Lower Pleistocene</td></tr>
<tr><td>600,000</td><td>Gunz-Glaciation
Donau-Gunz Interglacial</td><td>Tiglian</td><td>Basal Pleistocene</td></tr>
<tr><td>30,00,000</td><td>Donau Glaciations</td><td>Villafranchian</td><td></td></tr>
</table>

According to new dating techniques, Pliocene epoch began 13 million ± 2 million years earlier to Pleistocene. The Pliocene epoch continued till villafranchain stage. The basal Pleistocene began from 2.75 ± 0.75 million years and continued till 1 million ± 0.15 million years when Gunz-Mindel Interglacial period of lower Pleistocene began which was equated with Cromerian stage. This continued up to the end of Mindel glacial period. The middle Pleistocene began in Mindel-Riss Interglacial period about 0.3 million years ± 25000 years B.P. and continued till the end of Riss glacial period. The Riss glacial period began from 0.12 million years ± 5000 years B.P. and continued till 90,000 years ± 500 yeas B.P. The upper Pleistocene period began in the last Interglacial Riss-Wurm period by 90,000 ± 500 years B.P. and continued up to 8,300 ± 200 B.C when Holocene period began. This can be shown in the Table 5.2.

Table 5.2: Pleistocene Stratigraphy as per New Dating Techniques

Epoch	*Environment*	*Stage*	*Time in yeas B.P.*
Holocene	Post-glacial	Post-Pleistocene	8300 ± 200 B.C.
Pleistocene	Wurm Glacial Riss-Wurm Interglacial	Upper Pleistocene	90,000 ± 500
	Riss Glacial Mindel-Riss Interglacial	Middle Pleistocene	300,000 ± 2500
	Mindel Glacial Gunj-Mindel Interglacial	Lower Pleistocene	10,00,000 ± 1,50,000
	Gunz Glacial Donau-Gunz Interglacial Donau	Basal Pleistocene Villafranchian	2,750,000 ± 750,000

Ice-Age (Glaciation): As stated earlier, in the year 1909, Penck and Bruckner established Alpine-Glacial cycle. They established four glacial ages namely Gunz, Mindel, Riss and Wurm as followed by three Interglacials periods Gunj-Mindel, Mindel-Riss and Riss-Wurm. They discovered that second Interglacial period was longer than the other two Interglacial periods. That is why they named Second Interglacial period as the great Interglacial period.

With the discovery of Donau glaciation, Villafranchian fauna and Basal Pleistocene in recent years, we now have five glacial periods and four interglacial periods. The five glacial periods are Donau, Gunz, Mindel, Riss and Wurm. The four interglacial periods are Donau-Gunz, Gunz-Mindel, Mindel-Riss and Riss-Wurm.

The phenomenon of glacial may be described as increased precipitation in atmosphere which led to snow formations at high levels and accumulation at poles. The soft snow on accumulating started melting at lower levels due to pressure and solidified into ice. This simultaneously led to a viscous fore which moved the ice forward and away from zone of glaciations. Thus, large boulders and terrestrial materials were carried along with this moving mass of ice. The climate during this phenomenon was extremely cold and varied from dry to wet.

Causes of Glaciations: The main causes of glaciations are as follows:

- *Increased Temperature*: According to this theory cooling and increased moisture are pre-requisite for snow formation. A prolonged evaporation is responsible for both these requisites.

This is possible in presence of high atmospheric temperature. This hypothesis does not explain as to why snow formation takes place at mountain tops of tropical islands like Hawaii but does not occur in the plains of the same region.

- *Variation in Atmosphere Turbidity*: It has been observe that the average diameter of particles of volcanic ash is greater than the wave length of solar radiation and much less than that of terrestrial radiation. Hence, during volcanic activity solar heat was reflected back. But existing terrestrial heat was transmitted. This created a constant loss of heat in the planet. As a result glaciation was caused. Basically, arguments of this hypothesis are proven facts, the duration of such activities has a very short-term effect. Further glacial deposits do not show evidence of volcanic ash or similar particles of volcanic origin.
- *Polar Displacement*: This theory explains that the poles receiving minimum solar radiation are perennially covered with ice. Therefore, if there is a displacement in the whole or parts of the earth crust relative to its axis, then different regions will be brought in the positions of the shifted poles at various times. As a result of this, new areas start experiencing the conditions of the polar climate. This hypothesis has been rejected by scholars working in this areas because evidence for such polar displacement was very poor. The cyclic fluctuations of climatic conditions remain also unexplained.
- *Geometric Variation of Earth Orbit*: It has been discovered that there is a periodic shift in earth position once in every 21000 years. This affects the distribution of solar heat received, although the total amount of heat received remains constant. During this period, there remain long cold winters and short hot summers. Glaciations are explained as the effect of these long cold winters. However, geological evidences show that glacials and interglacials are longer in period than theory permits.
- *Solar Spot Theory*: According to this theory glaciations take place due to disturbance of heat equilibrium maintained in the planet. Disequilibrium is caused by periodic change of solar spot.

The solar constant of heat reaching the earth is 1.94 cal./sq. cm./ minute. Out of this 35 per cent is lost through direct and indirect radiation, while remaining 65 per cent is absorbed. Out of this absorbed heat 28 per cent is used up in evaporating water. Sooner or later the water vapour condenses in the atmosphere and releases heat of condensation. Thus, the energy deficiency of 28 per cent is balanced in the atmosphere. If this equilibrium between incoming and outgoing heat is maintained, the climatic condition will remain constant.

- *Planetary Motion*: Due to planetary motion, evaporation of water sheets on earth takes place. Through evaporation water is evaporated and reached in atmosphere. The blowing air brought them towards polar regions. High humidity and low temperature of polar regions turn vapours into ice. The ice-sheets were deposited at the top of Alps mountain.
- *Volcanic Incidence*: Volcanic incidence also causes evaporation of water sheets present on the earth. The evaporated water reaches in the polar regions where low temperature and high humidity turn them into ice.
- *High Temperature*: Planetary motion and volcanic incidence cause high temperature which causes evaporation of water sheet present on earth. The water vapours reach the polar regions through wind. They are solidified in the form of ice-sheet there due to low temperature.
- *Strong Air Circulation*: Strong air circulation is also responsible for the occurrence of Ice-age. The evaporation of water due to high temperature results in the formation of water vapours. Water vapours are blown towards polar regions where low temperature solidifies water vapours in the form of ice-sheet.
- *High Humidity*: Due to low temperature, there is high humidity in the polar regions. On cooling the water vapour results in formation of ice-sheet.
- *Low Temperature in Polar Regions*: Presence of low temperature in polar regions results in solidifying water vapours brought by strong circulation of winds in the form of ice-sheet.

All above mentioned causes were responsible for the occurrence of Ice-ages on the earth. Palaeo-climatological studies have proved that maritime climate was mainly responsible for the formation of ice-sheets. In the formation of ice-sheets maximum precipitation took place in the climate. When glaciation was in its maximal stage, it is known as maximum glaciation or glacial maxima. The water sheets on earth evaporated and precipitated in the form of ice on mountain peaks. The reasons for such maximum precipitation were planetary motion, volcanicity, high temperature, stronger air circulation and greater evaporation from the sea.

Consequences of Ice-Age. The consequence of Ice-age can be discussed under the following headings:

- *Formation of Ice-Sheet.* Ice-age had brought the incidence of formation of ice-sheet in the polar regions. During Pleistocene nearly one-third parts of the earth were covered by ice-sheet. When the glaciation was at its maximum, the ice-sheets had covered about 17 million square miles of earth surface, while at present they occupied about six million square miles.
- *Stages of Pleistocene.* On the basis of ice-ages, stages of Pleistocene have been discovered. These stages were namely Basal Pleistocene, Lower Pleistocene, Middle Pleistocene and Upper Pleistocene.
- *Stages of Glaciation.* The Ice-age created stages of glaciatioin namely, Donau, Gunz, Mindel, Riss and Wurm.
- *Stages of Interglaciation.* On the basis of stages of glaciation, stages of interglaciation have also come into existence.
- *Formation of Glaciers.* The Ice-age resulted in the formation of glaciers.
- *Formation of Moraine Deposits.* The Ice-age brought the evidence of formation of moraine deposits.
- *Formation of Loess Deposits.* The Ice-age was responsible for formation of loess deposits.
- *Stone-age.* The Ice-age brought the evidence of stone-age namely Palaeolithic, Mesolithic and Neolithic.
- *Stone-age Culture.* Ice-age gave the evidence of stone age culture, Palaeolithic, Mesolithic and Neolithic.

- *Change in Flora.* The Ice-age brought the evidence of changes in floral species.
- *Change in Fauna.* The Ice-age also gave us evidence of change in faunal species.
- *Pre-historic Man.* The Ice-age revealed us about the existence of pre-historic man and his culture.

Pleistocene Stratigraphy of Africa. When Europe witnessed the periods of glaciations and Interglaciations, Africa witnessed the periods of pluviations and Inter-pluviations during Pleistocene period. During Pleistocene, there was intense spread of ice-sheets and cold in temperate zones. But in tropical zones of Africa, there were heavy rains as indicated by river terraces. On the basis of study of these river terraces and evidences found in them, four pluvial and three Inter-pluvial periods have been recognised. Their names are as follows:

1st Pluvial	—	Kageran
1st Inter-pluvial	—	Kageran-Kamasian
2nd Pluvial	—	Kamasian
2nd Inter-pluvial	—	Kamasian-Kanjeran
3rd Pluvial	—	Kajeran
3rd Inter-pluvial	—	Kajeran-Gamblian
4th Pluvial	—	Gamblian period

The first pluvial Kageran is taken to be more or less contemporary with Villafranchian stage of Europe. During the second pluvial period Kamasian, eastern valley was almost entirely occupied by the huge lake Kamasia, the deposits of which are cut through by the famous Olduvai or Oldowan George.

Evidences of Pleistocene. The following are chief evidences of Pleistocene:

- *Moraine*: Moraines are deposits of objects brought by glaciers. This deposit contains stone pieces, tools, soil and other objects in the form of dune. It is believed that during the last two glaciations (Riss and Wurm), the climate of Northern Hemisphere was nearly four times colder than the present arctic region. Rocks and boulders were pushed along the tip of the advancing glaciers. These made mounds along the tip of the limit. There these glaciers stopped. Such deposits are termed as moraines. In this way moraines are the glacial features of melting ice which partly formed the basis for four-fold scheme of the glacial periods of Penk and Bruckner. They occur in glacial as well as inter-glacial periods. They play significant role in occurrence of glacial action, migration of life forms and also speak about altitude. As they are nearer to ice, there is remote possibility of their being associated with human habitation. Moraines are studied by geologists to understand the way and extension of past glacial phenomenon.

On the basis of position and place, the moraine can be classified into following types:

- Lateral Moraine
- Medial Moraine
- Bottom Moraine
- End Moraine

Lateral moraines are those moraines which are found along both banks of a glaciers.

Medial moraines are those moraines which are formed when two big ice pieces meet with force.

Bottom moraines are formed in the bottom of the glaciers.

End moraines are formed at the end point of the glaciers.

The phenomenon of moraine formation had taken place during Pleistocene. Hence, they bear testimony regarding the occurrence of Pleistocene.

- *Loess*: Dust particles carried away by wind and deposited at a place are called loess. The annual temperature during the peak glacial period went as far as down as—100°C. But during inter-glacial period the temperature did not rise beyond 0°C to 10°C. During the stay of ice, the groundwater was completely frozen (perma frost). As a result, water molten from the weight of ice could not get absorbed in the ground.

This created an expanse of slush around the lip of moraine. Advancing glaciers furrow through undulating land surface and in the process discharge a great deal of dust in the air. This dust is carried by strong wind current created due to the cooling of surface air. These wind borne dust particles are deposited everywhere. In the areas where glaciers could not reach due to intervening mountain, as in lower Australia, the airborne dust (called Loess) is found deposited to great heights. These loess deposits yield very rich assemblage of faunal material. Thus, Loess are composed of self-size particles ordinarily with accessory clay and sand and deposited primarily by wind. Loess is a sediment commonly non-stratified and commonly unconsolidated. The Loess that cover large parts of South-West Russia, East Europe and Middle Western United States have been derived from moraines and other glacial debris. Loess were deposited during glacial period, while soils were deposited in inter-glacial periods. As a result stratified sequences were built-up in Pleistocene. They were helpful in establishing Pleistocene chronology. The loess regions were grass land suitable for grazing animals like bison and mammoth. The hunters of these animals left their remains in composites located in loess. The soil horizons of loess form important markers of Pleistocene chronology.

There are differences of opinions among the scholars regarding the formation of loesses. It is believed that the climate was temporarily dry when they were formed during the Ice-age. In such conditions the sand particles from the surroundings of the rivers have been brought by the force of the wind and formed mounds. The fossil evidences of snail shells and mammals found in these loesses show that the climate was cold. In thickness, the loess range from 30 to 100 meters and are noticeable in France, Germany, Russia and Siberia. In North America the loess deposits occupy some thousands of kilometres. Spreading up to the Gulf of Mexico, an examination of these stratigraphy reveals the Ice-age to which they belong.

In France, the earliest deposits which yielded fossil remains of Villafranchian lines have been correlated with basal Pleistocene. In

South and Central Europe, loess deposits belonging to different periods have been found. They are known as older loess and younger loess. The older deposits were dark yellow brown in colour. They belonged to Sajee (Riss) glacial period. Due to climatic changes the upper portion became fossilised. Hence, it is called fossil soil. It is supposed to have formed when loess deposits stopped laying heaps of sand. In this interval there was vegetational growth. These were supposed to have formed during Riss-Wurm Inter-glacial period. The Loess deposits found over this soil is called younger loess. They are pale grey yellow in colour. They belong to Wurm glacial period. Over this layer, there are fossil soil deposits of two layers found in Central Europe, Czechoslovakia, Australia, etc. Thus during the ice-ages, the formation of ice-sheets was temporarily suspended to enable warm climate to prevail. Such periods are called Inter-stadial periods. If the change from cold to warm periods is of shorter duration, than that of inter-stadial, then the durations are described as oscillations. In the first three ice-age periods, there was a single inter-stadial period in each. But in the last ice-age period, there were two inter-stadials and three oscillations.

- *Changes in Fauna*: The end of Pliocene epoch and beginning of Pleistocene period was marked by the appearance of marine forceminifera species called Hyalineapaltrica which has been taken as a scale. The existence of this sea animal found in marine section of Calabaria at a place called Le Castella in Italy indicates the plio-pleistocene boundary. The Calabaria has been correlated with the mammalian fauna of Villafranchian stage. The animal species of Villafranchian age found in plio-pleistocene boundary are Elephas, Meridionalis, Dicerorhinus, Etruscus, Equus, Stenonis, Trogantherium Curverii and Dicerorhinus megarhinus.

By the Mindel glacial time, there were two new species called Dicrish Krichbergensis-merckii and a woodland variety called D. Hemitoechus. After Ester glacial period, a woolly rhinoceros species called Ceolondonta trickhorhinus antiquitais appeared in Asia and migrated to Europe as shown by the fossil evidence. During Mindel-Riss inter-glacial a deer species called Dama Clactoniana became extinct in Britain. In the last inter-glacial Riss-Wurm a deer species

Damadama appeared. By the end of last inter-glacial, the giant beaver Troganthérium became extinct. By this time, the musk ox also became extinct in Europe and reached cold climate. The other animals which became extinct are cave beer and giant deer called Megaceros gigantus.

The animals of Villafranchian stage in North Africa are Mastodon, Anancus Osiris, Elephas Africanavus, Stylohipparion lidicum, Equus namadicus, Libytherium marusium, Machairodus (Saber-toothed cats), Antelope, Gazello, etc.

In India in Siwalik strata animal fossil evidences were also found indicating plio-pleistocene boundary. On the basis of nature and fossil evidences, the Siwalik deposits are classified as Lower (Chinizi zone), Middle (Nagari and Dhokpathan zones) and Upper (lower tatrot zone). Siwaliks of Pliocene epoch. The animals of tatrot zone are Hipparion, Merycopotamus and proamphibos with Archidiskodonts, Camelus, Cervies, Hemibos, Sivatherium Leptobos, etc. By the time of formation of Pinjore zone Villafranchian stage, Hipparion, Merycopotamus, Proamphibos became extinct and Equus, Hypselephas, Rhinoceros, Bubalus species appeared. On the basis of extinction of animal species by the end of Tatrot zone and appearance of new animal species in the Pinjore zone of Villafranchian stage, the plio-pleistocene boundary in India has been recognised.

- *Soil Creep*: Soil creep reveals the deposits of stone particles on hill slopes. In glacial periods there were warm phases of short duration known as inter-stadial phase. Similarly during inter-glacial period also there were cold phases of short duration called stadial periods. In inter-stadial phase, the ice started melting due to rise in temperature. Water was formed due to melting of ice. Ice water formed in this way, started carrying stone particles. The current of ice water deposited stone pieces on hill slopes here and there. Soil creep bear testimony regarding the environmental change during Pleistocene period.
- *Changes in Plant Species*. Macroscopic and microscopic study of the plant remains found in lake sediments and bogs revealed that changes in plant species occurred with change

in the Pleistocene climate. The evidence of such deposits has been found on large scale in North-West Europe. The plant species such as Taxus baccata, trapantans, Vitis vinifera, Barsemia purpuria and Dulichium spathecium grow in warmer climate. Their presence in archaeological strata indicates the climate in which they grew.

The Donau period correlated with praetiglian phase was characterised by Azolla tegelensis, Tsuga, Najas intermedia, pterocarya limbergensis etc. Other phases recognised on the basis of floral evidences are Tiglian (inter-glacial), Eburonian (glacial), Waalean (inter-glacial) and Menopean (glacial); the last three equated with Alpine Gunz glacial period. The Cromerian stage has been correlated with the Gunz-Mindel inter-glacial period, where there is no equivalent phase to the Mindel glacial period. The Hoxian stage comparable to Mindel-Riss inter-glacial period has four sub-stages:

1. 1st sub-phase is Hippo phase which is characterised by the species of Pinus, Artemesia, Helianthemum, Plantago, Betula, etc.
2. Second sub-phase has a, b, c and d with the plant of Betula, Pinus, Quercus, Almes, Tilia, Azolia filiculoides, Ulanus, Carylus, etc.
3. Third sub-phase with temperate climate is second sub-phase.
4. The fourth sub-phase is glacial.

There is no comparable floral stage to Riss glacial period. The Riss-Wurm inter-glacial is correlated with Eemian stage during which there were Oak forets, Hazel, Horn beam, etc. Basing on plant evidences Wurm is divided into early wurm, main wurm and late wurm. The early wurm is characterised by the plant species like Betula and Pubescences, Populus Trimula, Pinus silvestrus, etc. The main wurm is characterised by dry weather and one oscillation called Paudorf inter-stadial oscillation. The late wurm has three zones:

Zone I. It is characterised by (a) oldest Drayas phase and Arctic flora, (b) boilling oscillation and temperate plants and (c) older Drayas and Alpines.

Zone II. It is characterised by Allerod oscillation and Pines.

Zone III. It is characterised by younger Drayas and cold climatic conditions.

Zones IV to VI. These zones are characterised by pre-boreal, Boreal and Atlantic with Tundra forests; Pine, Hazel and Oak tree and oak and elm forests, all belonging to post-pleistocene stages.

- *Terraces.* Terraces represent the life history of a river. They are former flood plains formed along streams due to the changing course of the river. These involve periods of aggradation or deposition and degradation or erosion.

In the period of aggradation the river deposits silt and gravel and builds up its bed. But in the period of degradation the river cuts its bed. A number of rivers of the world present a succession of terraces which usually contain the remains of animals and men who lived on the river banks during the formation of their respective deposits. The study of these remains has an important bearing on the chronology and climate of early man and his cultures.

The river terraces are the result of the processes of valley silting and valley erosion. They were formed due to changes in sea level and changes in the volume of river. The active erosion of a river till the slope of its bed or longitudinal profiles acquires a form known as curves of equilibrium. This is nearly parabolic step near the source and flat towards the mouth. Here rivers meanders on its flood plains. Erosion in upper stream and the deposition in the lower reaches by this stage become balanced and minimal. The river is now graded as base level. The base level is the level of body of water below which erosion of a river cannot take place. Any change in base level affects the activity of a river. In this way lowering of base level is brought by erosion and rising is effected by deposition of its bed.

The Pleistocene epoch witnessed the glaciation in Europe, America, Asia and Africa simultaneously. Terraces formed in the lower reaches of the rivers were mainly due to the periodic withdrawal of water from oceans as ice sheet grew on land. During the glacial period, there was an active erosion of the riverbeds in response to the lower base level of the sea which was in regression. The erosion which starts at mouth level works up stream. The coming of inter-glacial period makes the sea level rise again due to the melting of ice-sheet. Now the rivers become sluggish, silt up new channels and build up thick deposits on their valley floors. This process is known as aggradation. The changes in the sea-level is not recorded immediately in the upper course of the rivers. Each change goes up slowly.

Alterations in the volume of water and sedimentary load due to climatic changes were the main factors responsible for the erosion and aggradation in the middle and upper portions of a number of European rivers during Pleistocene times. It is also clear that aggradation was characteristics of inter-glacial period near the mouth of rivers and degradation was characteristics of glacial times. The higher river terraces are mostly older. All the main terraces run into ancient sea levels and occur in tiers on the raised beaches. All these reflects the rise of land masses throughout quaternary times. Though as a rule the stratified deposits on a low terrace are younger than those on a higher one, a low terrace sometimes include portions of a deposit laid down in deep channel broadly contemporary with or older than the deposits formed on higher terrace.

- *Sea Levels*: There are evidences to show the occurrence of sea level changes during the Pleistocene epoch. During the glacial periods the water from the sea evaporated and precipitated in the form of ice at higher levels resulting in reduced or lowered sea level where as during the inter-glacial times, the ice sheets melted and flew into rivers and sea leading to the formation of raised sea levels. In this way, changes in levels of seas occur due to direct expansion and contraction of Pleistocene ice sheets. The relationship between sea levels and glaciation is complex for various reasons. But there are two main factors: Eustatic factor and Isostatic factor. Eustatic has affected all the oceans and open seas of the world. This factor caused alternative withdrawal and release of water due to formation and melting of ice-sheets. Isostatic factor has effected the earth's fall and rise under the weight of ice which was restricted to areas subjected to glaciation. Eustatic movements are independent of Isostatic movements or movement of earth. But if there were no movements of earth, there would have been no evidence of lowered sea level available for observation. The land masses depressed under the weight of ice had raised the level of land when the ice retreated. Thus, the Isostatic movements had raised the land in some places to greater heights above sea levels as a result of which traces of old, once more lower shore line are visibles. In the areas unaffected by glaciations the tectonic movements irrespective of whether they were covered by ice or not have exposed Pleistocene shore lines.

The alternating lowering and raising of the sea resulted in changes in sea level of at least 300 feet. There are evidences to show that in some inter-glacial periods the sea levels were 200 meters higher than the present ones while in the glacial times they were 150 meters less than the present ones. The present heights above sea level of fossil shore lines or raised beaches have been largely used as means of relative dating. But such attempts have been reported to be difficult or unreliable in regions affected by local depressions of land due to Isostatic movements. However, sea level fluctuations have left useful records for archaeological interpretations when the archaeological materials are found in association with datable geological phenomenon.

The Pleistocene sea levels could be recognised on the basis of ancient raised beaches and fossil animal remains found in them. A good deal of research of these phenomenon has been carried out in the Baltic and Mediterranean. On the basis of study of ancient raised beaches on the Mediterranean coast seven ancient raised beaches ranging from an altitude of 200 meters above present sea level have been observed. These stages are known as Calabrian stage. They have been correlated with basal pleistocee. Calabrian with an altitude of 200 meters above present sea level, Skilian with an altitude of 39-100 meters, Milazian (Gunz-Mindel inter-glacial) with an altitude of 55-60 meters, Tyrrhenian (Mindel-Riss) inter-glacial with 28-30 meters altitude, Monastrian I with altitude of 18-20 meters, Monastrain II with altitude of 6-8 meters and Epimonastrian with an altitude of 3 meters. The Monastrian II stage has been dated to above 80,000 years with the help of uranium-thorium method of absolute dating. Ancient raised beaches have also been observed in North Africa, South France, South England and East North America.

6

Evolution of Man and Conquer Over Nature

When and how man emerged on the earth is still a question to solve. Divinal theory explains about the creation of man by God. But evolutionary theory postulates the origin of man from simple to complex form. Today there are fairly a good number of people who are advocating till recently the origin of man from as diverse source as Dinosaur at one hand and life from other planets on the other. Those who had been following the organic evolution are also not in agreement about as to how this process gave rise to man and when. Evolutionists believe that man is a member of primate family which includes monkey; anthropoid apes and hominids. Therefore, ancestors of man, hominid, anthropoid ape and monkey were the same. These ancestors would have emerged from mammalian animals which would have emerged from other creatures of Proterozoic age: Among different non-human animals such sequences of evolution have been established. For example, horse and camel. But unfortunately we do not have sufficient evidences to prove orderly evolutionary sequences of man. So, we are facing the problem of missing link. The missing link is still unknown. But absence of missing link cannot be taken as the proof that evolution is a faultful theory. As we have seen earlier that main means of studying evolution of animals is sedimentary rocks. In sedimentary rocks, fossils of those animals are found which were drowned in water. But early man did not know to swim so he did not attempt to go and swim in seas. That is why fossils of man are not found in sedimentary rocks. Another problem is sedimentary rocks situated in different countries of the world have not been studied and researched properly. Therefore, we do not find sufficient evidence of evolution of man.

Early Ancestor of Man

Who was the early ancestor of man? We have a number of misconceptions in this regard. Generally it is believed that evolutionists consider monkey or ape as the early ancestor of man. But this does not hold truth. Evolutionists have considered the evolution of man from not ape but from some Anthropoid apes. According to them Anthropoid apes are tailless like man. Their blood also resembles with human blood. The disease of Chimpanzee can spread in man and the disease of man can spread in Chimpanzee. The ovary of females of Anthropoid apes resembles with the ovary of women. The life span of man and Anthropoid ape is also nearly similar. These characteristics of man do not resemble with other apes. That is why evolutionists have attempted to show the evolution of man from Anthropoid apes. But there is difference of opinions among the scholars regarding who this Anthropoid ape was? Today four types of Anthropoid apes are known to us. They are Gibbon, Chimpanzee, Gorilla and Orang-utan.

Fig. 6.1: An Imagination of Evolution of Man

Some scholars believe that man would have emerged from two or three types of Anthropoid apes. For example, Negro of Africa from Gorilloid animals, Chinese from Orang-utanoid animals and white races of Europe from Chimpanzeoid animals. But it has been established today that all human beings have evolved from the same

Fig. 6.2: Dryopithecus

ancestor. According to scientists, this animal was Anthropoid ape but different from Gibbon, Gorilla, Chimpanzee and Orang-utan. The brain capacities of these four Anthropoid apes are so low that they could not be considered as the ancestor of man. Who this Anthropoid ape as ancestor of man was? Where and when this animal originated? These questions are still left to answer by the evolutionists. Probably this animal would have possessed bipedal walk and lived on land but was able to climb on tree easily. Their forefathers would have left climbing on the tree in Cainozoic Age. His cranial capacity would have been greater than Anthropoid apes. The head would have been like club-shaped. The nose would have been not flat like monkey, rather nose would have been elevated. Hairs would have been less on body than apes. This unknown animals is known as missing link of human evolution. In discovering the missing link, scholars have been able to discover a number of fossils of hominid animals. These hominids resemble with man to a great extent. The discovery of these hominid fossils have not solved the problem of missing link but has been able to find out several animals who were very similar to man but were not Homosapiens. They were hominid. The writers of first chapter of culture were not Homosapiens rather they were Hominid. These Hominids proved that man is superior to other animals and possessed efficiency for making tools and culture. By making tools and culture, man became conqueror of nature.

Clues of Man's Conquer Over Nature

Man is a mammalian animal. His young ones have to depend upon parents for a long time for their protection and survival. His young ones are socialised not only in family but in community too. Socialisation process provides children to acquire the experiences of community members. They develop social virtues by which they live as a member of society. The young ones of other animals lack these social experiences. Other animals lack these social experiences. Other animals have to struggle independently for survival. They have to move on the basis of their own experience But facilities of social experiences were available to all mammalians to some extent. The number of man was very less in early time. Man was not stronger than lion, tiger, bear, elephant, etc. But why he became conqueror of the nature and established rule over other animal?

Man could become conqueror over nature and established rule over other animals because he developed efficiency to adapt himself with changed environment. The nature provided him such ability by which he could rule over the trouble. He possessed efficiency to make discovery and invent tools to cope with changed environment. His key to success over nature are as follows:

1. Speaking power
2. Expression power
3. Tool making

Speaking power: Oration power of man is superior to other animals. He can pronounciate varieties of sounds. Other non-human animals also possess this quality. But they can pronounciate very limited sound symbols. Man is such an animal who can pronounciate varieties of sound symbols by subtracting and adding phonemes. He is also able to provide social meaning to his sound symbols. When human children know to speak, they know the social meaning of the sound symbols of the community. They make conversation with community members easily through the sound symbols of the community. This is also known as dialect and language. Dialect or language provides occasion to acquire the knowledge of the community. Man is in a position to tell his children as to how he can save himself from lion, tiger, bear, elephant, fire, flood, etc. Languageless or dialectless other animals are not in a position to educate knowledge in their children.

Expression power: Due to possession of language or dialect, man is in a position to express his opinion. He can make support or express disagreement through expression power of language or dialect. Man can express his opinions on different topics. For example, fire can destroy cloth, body, house, property, etc. Flood can cause great damage. Dirty water brings disease. Keep your environment clean, etc. Other non-human animals lack expression power. The power of expression is such a valuable thing by which man gets success over troubles.

Creation of Tool: The hands of man were used to carry the weight of his body in early time. But when he began to stand erect and walk, his hands became free from carrying the weight of body. When the hands of man became free, he began to perform different activities. The tools of other animals such as palm, nail, beak, etc. are attached with their body and can perform limited activities. But man possess opposable and movable thumb. The structures of toes and fingers of man are quite different from other animals. These are helpful in making tools of different types and perform varieties of activity like gathering, collecting, producing, cooking, protecting, decorating, playing games and music, writing etc. This facility is not available to other animals.

Major Stages of Human Evolution

Scientists working through fossils felt that the separation of apes from the line which gave rise to man took place somewhere between 20-30 million years ago. On the basis of study of albumens, some molecular biologists have recently shown that this bifurcation may have been as late as 5 million years ago. The disagreements about human evolution are too many and can be presented in a book separately. Hence, we are going to describe only four major stages through which human evolution has passed.

Ramapithecus: 10-14 million years old geological layers have yielded some 40 teeth and some 15 fragments of jaws from India, Kenya, Pakistan, China and other regions of the old world. Since no other parts of their body is known we have very little ideas about their height and walking posture. The most important feature which led them to be our direct ancestor is extreme reduction of their canines and modifications of chewing teeth. These evolutionary changes are impossible unless the hands have been reasonably free. Ramapithecus believed to have already entered into the direction of specialisation that leads to the human kind (humanisation process). The evolutionary significance of freeing of hands has to be understood at this junction.

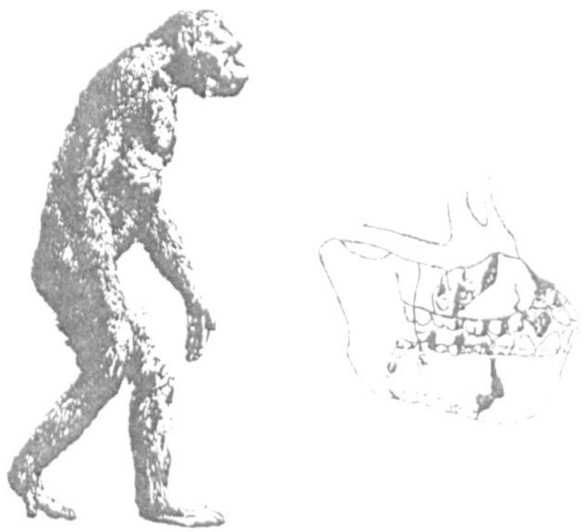

Fig. 6.3: Ramapithecus

Most of the physical anthropologists believe that a group of brachiating arboreal apes had come down to ground for terrestrial living. The ground living required a different kind of adaptation. It is believed that freeing of forelimbs is related to ground living adaptation. Further, it is argued that the nuts and fruits growing on ground being harder than those growing on trees. In this situation these apes had to develop strong chewing apparatus. The long canines are not very useful for this purpose. The chewing requires a great deal of side to side or rotating motions of the jaws. The canines of two jaws in the case of apes were often interlocked and hence were a great hindrance to this adaptation. It is quite likely that Ramapithecus represents one of the first steps of adaptive process.

Ground living required not only the adaptation of chewing apparatus, but also combating with an entirely different set of dangers which were non-existence in tree living. It is believed that culmination of all these précising needs brought about a development of bipedalism. Walking erect made man lose the snout, free his hands, balance his head on spinal column by developing occipital lobe of the brain and finally ability to manipulate the hands to work with the environment. Thus, a chain of changes took place in apes when they came down to live on ground. It is believed that by about 5 million years from today he was capable of breaking simple natural objects like tree branches or stones to use them for his defence and occasional offence. The meat eating was limited to carcass eating from the left over of other predators.

The recent discovery of about 8 million years old primate skull, which is much more complete than previously known fossils of this period, is more closely related to Orang-utan and cannot be a human ancestor.

Australopithecus (Afarensis): Australopithecus was a light, gracile creature who stood a little more than 4 feet in height and walked erect with a curved back. This form was a more suitably adapted to ground living and survived during 4 millions to a million years from today. His brain was larger than present day Chimpanzee and his teeth were still in the process of evolving. A group among them seem to specialise in a different diet and developed a barrel like trunk with an altogether robust built. These are identified as forming a separate evolutionary specialisation and are called Australopithecus robustus. Both these species became extinct before 1 million years.

Fig. 6.4: Australopithecus Africanus

Recent evidences from Africa have revealed big colonies of Australopithecus with piles of eaten bones and numerous stone fragments. These evidences leave little doubt about their ability to shape some sharp boarders by simple breaking of stones and using these sharp boarder stone fragments as tool to cut animals. Most probably they still did not hunt but merely scavenged dead animals. There is increasing belief that Australopithecus co-existed with another advance variety of ape man who formed the first member of hominid bifurcation branch. The eastern rift valley in Africa continues northwards through Ethiopian highlands. Don Johanson in 1973 discovered an almost 40 per cent complete skeleton from these deposits at Hadar. The deposit could be dated to over 3 million years and the skeleton was named Lucy. The find was undoubtedly one of earliest species of Australopithecus which may have been the first representative of hominid bifurcation from Australopithecus. It was given the generic status Australopithecus Afarensis.

At Koki Fora region around lake Turkana another group of finds dating to 1.8 to 1.6 million years alongwith many tools and Australopithecus remains were described since late 1960s. In 1961 Louis Leakey had announced the discovery in Bed I, Olduvai Gorge of a more modern looking hominid, which he called Homohabilis or the hand using man. Thus, the earliest membership in the hominisation process that had to be ascribed within our own genus was given to this fossil find. At Koki Fora Homohabilis could be attributed with more of tool making ability on added evidence. In other words, the direct ancestor of Homo erectus had to be now considered as Homohabilis rather than Australopithecus. Whenever Habilis is reported, it is invariably accompanied by a large number of Australopithecus boisei finds. Further, it is important to note that there are specialists who would still like to group Habilis with a somewhat advanced variety of gracile Australopithecus. Therefore, latter group of specialists would find it not difficult to accept that tool fabrication and manipulation emerged within Australopithecus stage.

Fig. 6.5: Java Man (Pithecanthropus Erectus)

Fig. 6.6: Sinanthropus or Chinese Man (Pithecanthropus Pekinensis)

Homo Erectus: Homo erectus was much closer to man in looks. His head was more rounded and had a volume of nearly 1225 CC (modern man 1500 CC and ape 450-600 CC). He was nearly 5 feet in height and had strong bones. The earliest of them has been found from lake Turkana in Kenya and dates around 1.5 million years. The original finds from Java (Pithecanthropus) and China (Sianthropus) were of much younger date (200,000 to 400,000 years). This demonstrates the long period through which they ruled the forests. Erectus was perfectly adapted to moving varieties of stonetool, the magnus opus of their creation being handaxe. More than 60 species of animals have been found associated with these tools. These include elephant, rhinoceros, bears, horses and camels beside numerous smaller species. Sometimes in the long duration of their stay, erectus must have demonstrated fire because we have evidences of hearths where they must have kept the fire alive. It is believed that big game hunting was possible mainly because of domestication of fire. Herds of these large mammals could be driven into swamps or driven off a cliff with the use of fire and once trapped, they could be butchered.

Early Homo sapiens: When and how these fairly well equipped Homo erectus evolved into a group of early Homo sapiens is difficult to demonstrate. The generally held view is that these early precursors of modern man might have evolved between 200,000 to 100,000 years ago in Africa, Europe and Asia. They show no specific or radical change in their life style because they are also found with the same hand axe tool kit of erectus, albeit much more refined. Their physical features are, however, much changed with more strengthening of chewing muscles and thus, putting out of temporal lobes of the brain (above the ears). This added an array of additional brain faculties. All authorities agree that language must have already developed by the time of erectus evolution. Therefore, early sapiens were equipped with that we are biological endowed with. They are known as Swanscombe man, Steinheim man and Fontechevade man in Europe.

Here, it is significant to look into the impact of biological evolution on the life style of our ancestors during this period (600,000 to 100,000 years before). Standing erect required a modification of pelvis which narrowed the birth canal. The growing brain box and narrowing of birth canal were two evolutionary changes that took place simultaneously. These are inferred with each other in the birth of human child. Nature brought a compromise in this conflict by allowing a baby to be born much before it develops a full brain size. This required lengthening the period of post-natal dependence of new born. Evolution had also taken away the phenomenon of estrus in this group sometimes during this period. This made the female sexually receptive all round the year. These two combined to make human group ties stronger and stabler. Whether an ideal family could develop in early Homo sapiens period is difficult to prove but they were able to organise in to stable group.

Neanderthals: Neanderthals are believed to be a specialised branch of early human races who evolved from erectus stage around 100,000 years ago and lived simultaneously with early sapiens for nearly 65,000 years. This race was a branch of main line of evolution. It is mainly believed because his brain capacity evolved to a limit which exceeded that of modern man by 50-100 CC. In spite of his large head his other physical characters remained fairly rugged. He had large cheekbones and a very prominent brow ridge curving over the eyes and connecting across the nose giving rise to a deep sunken appearance to the nose. These features were further accentuated by rather a broad nose which was perhaps an adaptation to cold climate.

Coupled with these facial features, his thick set body with a slight bent and a height of 5 feet gave him a real demonic look. Why such a form had to evolve when the generalised trend in evolution was progressing in a different direction is evident from the fact that they were the first proto-human groups who penetrated deep into north arctic climate. In terms of their geographical spread also the Neanderthals excel all other contemporary or earlier forms.

Regarding the socio-cultural life, the Neanderthals show some very interesting evidences. Their tool kit demonstrates a very specialised series of relatively more job specific flake tools which were successfully used in large game hunting. They are also the first among our ancestors who show evidence of pre-occupation with life after death. They used to bury their deads with food and tools laid along with the body. Another evidence of their skill and emotions comes from Shanidar cave in Iraq. Here a 30 years old man has been buried with flowers covering his body. This is a sentiment not quite uncommon even today. This man had one of his arms successfully amputed during his life time. Many other Neanderthal burials are known with evidences of religious cult, connected with cave bears. In one instance, an arthritic old man is buried which clearly shows that even crippled people must have been adequately cared for by them. All these evidences qualify the Neanderthals as very close to us in their attitude and sentiments. Their physical ability and degree of environmental adaptations was also no less than what was required for a successful survival and yet this race became extinct by 35,000 B.C. with the advent of modern man. A possible explanation given for abrupt extinction of Neanderthals is a direct confrontation or bloody war between these two races. So far, we do not have any archaeological evidence to prove this possibility. But they were eventually wiped out of this planet seems certain.

In the sequel, it is important to mention here that there are some indications of interbreeding between the Neanderthals and modern man in cave sites in Yugoslavia and Iraq. It is possible that these inter-mixtures have finally given rise to upper Palaeolithic modern man, Homo sapiens sapiens.

Homosapiens Sapiens (Modern Man): Homosapiens or modern man appears rather in an abrupt manner in Europe around 36,000 years ago.

Right from the stage of early appearance it shows rather strong internal heterogeneity of form. Many scholars believe that this heterogeneity may have finally given rise to three main racial groups of modern man. These racial groups were Cro-Magnon, Grimeldi and Chancelade man. Cro-Magnon was one of the most talked forms of modern man found from the rock shelter of Cro-Magnon in France. He is often taken as the ancestors of modern Europeans.

The sites of occupation of this race of modern man are invariably very thick in their occupational debris. This is taken to indicate a regular population growth among them. They not only migrated all over Europe, but their habit of migrating seasonally with their prey made them enter land which were not inhabited by our earlier ancestors. Thus, people of new world took place in this period. They must have also devised ways to navigate and float. Otherwise they would not have been able to migrate Pacific Islands and finally to Australia.

The isolated skull found from Niah cave from Borneo can be taken as a direct evidence of this migration to South-East. There are, however, some very recent evidences coming up from both Australia and North America which show that these migrations may have taken place in several waves. The earliest waves of the migration to New world may have been some Neanderthal races, not modern man.

The true Homosapiens or modern man was not only adapting rapidly to various environmental conditions. But his extreme mobility made him develop sharply different cultural habits in different parts of the world. Out of the most significant of his cultural achievements was his extreme pre-occupation with visual art. He under took great pains to draw and paint various figures and motives in hard rock walls, ceilings and floors. The motive of his art can be taken as creations for more visual pleasure because there are several enigmatic features about them which can be explained at an esoteric level.

Race Differentiation: Modern scientific researches have demonstrated that human adaptation within these widely varying environments must have already started the genetic selection process. For instance Vitamin D synthesized from solar radiation is both essential as also lethal when taken in large dose. Higher exposure to sun could also be lethal and hence melanin (Pigment causing chemicals) of the skin is activated to protect the body from absorbing excessive Vitamin D. That is, all human being have the same number of pigment producing cells (melanocytes), the rate of pigment production is differently programmed by genetic selection.

Similarly, mutation may have caused sickling phenomenon of blood just in order to enable man to adjust with an area otherwise infested by malaria. We have very little possibility of knowing through archaeological researches, how and when the various differentiating process in man started. It is generally believed that man's need to adapt to various kind of environmental and cultural stress faced during the Pleistocene period may have been the main cause of evolution and subsequent differentiation.

Coprolite or dried faces of pre-historic man have been analysed to see the pattern of food man has been eating. Many interesting informations regarding his preference of specific edible plants out of a large available variety (known through polynology) can be demonstrated through this study.

Similarly, palaeo pathologists, through the study of skeletal remains, have demonstrated that arthritis was perhaps one of the most common ailments suffered by pre-historic people. For comparatively recent periods, evidences of many other diseases which have a permanent mark on bone and teeth have also been recorded.

Main Ages of Human Civilisation

We have seen earlier that the main difference between man and other animal is that man builds tools and implements for his protection and safety. Other animals have only beak, trunk, teeth, horn, tail, etc. which they use as tool for protection. Then these tools of protection are attached to their body. But man has been producing tool from early ages. In the early age he used small branches of trees and stone pieces as tools. Thus, he was not producer of tools. Rather he was user of tools provided by the nature, i.e. forest and mountains. On the basis of his experiences with nature, man gradually learnt to fashion tools from stone pieces. In the early age his tools would have been big and very simple. But with the passage of time, his tools became more and more small, beautiful and useful. Thus, the tools made by man were sign of his mechanical, industrial and scientific success. Archaeological anthropologists have divided the human history into different ages on the basis of different materials used by pre-historic man for the production of his tools. The two major division of human history on the basis of materials used for shaping tools by early man are as follows:

1. Stone Age
2. Metal Age

These two ages have been divided into small periods for the purpose of study.

Stone Age: The first age of human civilization is known as stone age. It is because in this age the tools and implements made by man were of stone. This age runs parallel to Pleistocene age. It covers nearly 99 per cent parts of human history. The tools and weapons fashioned by man in the early time of his existence look like natural stone pieces. These tools are known as Eoliths. This age is known as Eolithic age and the culture is called Eolithic culture.

From the period of 1st Interglacial one finds the existence of such tools which can be easily accepted as made by man. Such tools have been divided into following ages:

Paleolithic Age: Time span of this age has been established as from 5-6 lakh to 12 thousand years from present days. The livelihood of man of this age was dependent upon hunting and gathering. Animal husbandry and agriculture were not known to him. This age is divided into three periods namely, lower Paleolithic, middle Paleolithic and upper Paleolithic periods. Core tools are characteristic tools of lower Paleolithic period. They are handaxes.

The tools of lower Paleolithic period were big and ugly. The tools of middle Paleolithic period were definitely smaller, beautiful and more useful. They were flake tools. The tools of upper Paleolithic period were made of blade and bone. They were more in number and varied technologically. Scraper is characteristic tool of middle Paleolithic period. Blade and burin are characteristic tools or upper Paleolithic period.

Mesolithic Age: The time span of this period is taken as between the end of Pleistocene and beginning of Holocene. This is taken as a transition phase. The tools of this period were microliths. Bow and arrow are characteristic tools of this age alongwith microliths.

Neolithic Age: Geological period Holocene started nearly 10 thousands years ago. The later part of Holocene period is taken as Neolithic age. The characteristic tools of this age are Celt of different types. They were made by polishing and grinding techniques. Man had acquired the knowledge of animal husbandry, horticulture and agriculture in this period.

Metal Age: Metal age started 6-7 thousand years ago in Indus Valley, Egypt and Creaste. This age can also be divided into following smaller ages namely Copper Age, Bronze Age and Iron Age.

Copper Age: In the early 2000 years of metal age man lived in copper age. Besides the use of copper tools, stone tools were also used in this age. That is why this age is also known as Chalcolithic age. In this age boat, wheel and potter wheel were invented. Bullock carts were also made. The concept of specialist and special class developed in society.

Bronze Age: In the last part of copper age, man new to mix tin in brass. As a result of mixture of these two metals bronze was invented. The time span of this period is 4,000 to 2,000 B.C. The tools made from bronze were hard and more efficient. Trader class came into existence for export and import of metals. Man began to settle in fertile valley. He constructed dam on river and built canal for the purpose of irrigation. Irrigational civilization, complex society and urban centers came into existence. Army, law and court were established. Writing was also invented to maintain record of trade and payment. Coins were invented. In this way urbanization came into existence.

Iron Age: This age came into existence in 1500-1200 B.C. Iron tools were more efficient than bronze tools. These were more suitable for agriculture.

From iron age, man entered into steel age and plastic age. Now we are living in electronic, science and technology age.

Table 6.1: Geological and Cultural Chronology

Geological Periods	*Time*	*Culture*	*Human Species*
Holocene	1000 B.C. 4000 B.C. 6000 B.C. 8000 B.C. 10000 B.C.	Iron Age Bronze Age Chalcholithic Age Neolithic Age Mesolithic Age	Modern Man Modern Man Modern Man Modern Man Modern Man
Upper Pleistocene	Wurm Glaciation (100,000 BP)	Upper Palaeolithic	Homo sapiens, Cro-Magnon, Chance lade man, Grimaldi man
	Riss-Wurm Interglaciation	Middle Palaeolithic	Neanderthal Man
Middle Pleistocene	Riss Glaciation (200,000 BP Mindel-Riss Interglaciation	Lower Palaeolithic	Same Human Species
Lower Pleistocene	Mindel Glaciation (400,000 BP Gunz-Mindel Inter Glaciation	Lower Palaeolithic	Australopithecus Homohabilis, Homo erectus, Early Homosapeins
Basal Pleistocene	Gunz Glaciation (600,000 BP)	Eolithic	Australopithecus
	Donau-Gunz Interglacial Villafranchian or Donau Glaciation (3000,000 BP)	No Known Human Culture	No Known Human Species.

7

Evolution of Eolithic Man and Culture

Early Tools

The greatest problem faced by the earliest man was his protection from wild animals and collection of food materials. His number in comparison to other wild animals was very less. He was not in a position to compete physically with them. Physically he was very weak in comparison to lion, tiger, wolf, elephant, bear, etc. But man was superior to other animals in the construction of his hands. He possessed freely opposable thumbs and fingers which helped him in picking up soil pieces, stone pieces and wood pieces. In the beginning, he used these naturally broken things as tools for his protection and food collection. Even today we use them for plucking fruits and beating animals. But remains of these tools used by early man are not available to us because they were perishable things. These things were provided to man by the nature. They were not produced by man. Rather they were used by man only. They did not possess human activities. Hence, they cannot be placed under the category of tools. If the earliest man would have used some activities in making them, we are quite unable to tell anything with certainty because we have no any evidence as such till date. But we do not find such matter related to stone because stone pieces are durable. Their remains could be preserved by the nature. Man was aware of their utility since beginning. Probably he would have used some naturally broken stone pieces for hunting small animals and taking out skin hide. He used to find such stone pieces nearby his habitat. These stone pieces were not pointed. They did not possess sharp edge. But when man acquired the technique of breaking stone pieces further and making sharp edges by striking against another stones or hammering by another stones;

progress of the earliest man in this direction became easy. Now man was able to understand that such tools were useful for not only hunting and gathering but also for various other activities.

The Problem of Eoliths

We cannot say with certainty that the earliest man was knowing to fashion tools in beautiful manner. He had learnt this technique in several million years. The stone tools fashioned by him were very much similar to naturally broken stone pieces. But some chips were removed to get a desired edge. It is very difficult to distinguish between tools and naturally broken stone pieces. This is a matter of debate among archaeologists, too. Some treat them as man made, while others consider them as natural stone pieces. The time period of these stone tools have been established as the later part of Pliocene to 1st interglacial age. Archaeological anthropologists call these tools as Eolith or Downstone. The age in which these stone tools were made is known as Eolithic Age.

The Eoliths are alleged cultural objects fashioned by human handiwork which have been reported time to time from pre-Pleistocene age. The first Eolith was discovered by Abbe Bourgeois in 1867 in an assemblage of chipped flints in the bed of upper Oligocene age near Thenay, a village in South Osteans. In 1877, similar objects of upper Miocene were reported from France. The reality of existence of these alleged man made cultural objects has given rise to a prolonged controversy. The main difficulty in outright rejection of these reports of culture was lack of any conclusive proof that these were shaped by natural process. Even today we experience the difficulty to distinguish between art and nature or between man's workmanship or shaping of stones by natural causes. This is specially a great hindrance in the identification of extremely primetime cultures. Believers of Eoliths regard the presence of bulb of percussion, pointed ends, curvilinear notches, the so-called resolved flakes and the marginal retouches as indicative of human workmanship. Reid Moir thought that the distinction lies in the angle at which flakes were removed. Eoliths are characterized by one sided flaking, or a dom or crested top, a flat and naturally fractured base and a deep edge flaking made by blows delivered around the under surface.

Rostro-Carinates

The problem of Eolith was studied subsequently by Reid Moir, especially in the context of crag deposits from the stone bed in the Ipswich, an area in Norway. Those were datable to closing part of Pliocene. Occasional quartz and quartzite module flakes have been found with animal fossils. The bed has yielded awls, scrappers, a few hand axes and abundant flaked implements in either disturbed or undisturbed position. Among those rough primitive flake implement, the rostro-carinate is typical. Laknester has termed as Icenian industry. Stamp and Hamer have designated them Ipsvician. Implements are of a rich mahogany colour. They resemble with the shape of an upturned boat. They are shaped from nodules of flints and limestone. They have resulted from a crude attempt to produce point and edges. Because of their edge chiping and surface flaking, they have been placed midway between Eoliths and Palaeoliths. They have been taken to be in direct ancestral line of more elaborate traditions like Chellean and Abbevillian.

Eoliths have been reported from Britain, France and Germany in Europe. They have also been reported from Egypt, Russia, Burma and Australia. Such industries of crude form are datable from Eocene to Pliocene. It is significant to mention here that such Eoliths are also described from post-glacial Pleistocene as well. Based on various kinds of flints obtained from different geological layers, M. Rutot has described six phases of Eoliths, which are as follows:

	Phase of Eoliths	*Geological Layers*
1.	Thenetian	Lower Eocene
2.	Faugenian	Middle Oligocene
3.	Thetrayan	Upper Oligocene
4.	Cantalian	Upper Miocene
5.	Kentian	Middle Pliocene
6.	St. Prestian	Upper Pliocene

Some scholars do not accept Eoliths as pre-palaeolithic in age. According to them, it represents a primitive type which survived alongwith palaeoliths. While admitting that Eoliths represent infancy of stone cultures, some scholars go to the extent of formulating a pre-eolithic stage, which is only hypothetical in nature. This period of culture called Alithic had probably wood implements or merely teeth, claws or natural picked up pebbles.

The primitive method of stone flaking indicates that an Eolithic stage in the evolution of stone implements is unlikely. As they are usually found overlying gravel deposits, the critics of Eolithic existence infer them to be formed out of pressure and movements. Even the finest flaking, the notches, parallel chipped edges and bulb of percussion taken as international working are claimed to be possible results of sea waves of swirling streams or heavy treads of camel and other animals (podoliths). Their rigid unchangability from the lowest tertiary to the present and their constant repetition to form suggest natural which do not vary in their operation. Although Eoliths have been reported abundantly from various places of different ages, they have never been found associated with human occupation. Thus, on the basis of typological, environmental and palaeontological grounds Eoliths appear to be most improbable.

It should be relevant to mention here that Eolithic claim was linked up with a presently discarded concept of human evolution held by many during the early part of this century. This is about the theory of two parallel phyla in to which the human line bifurcated in its course of evolution. The one which eventually led to the modern races was termed Neoanthropine branch while the one which became extinct was termed as Palaeanthropine branch. Thus, it was convenient to explain the simultaneous occurrence of primitive and advanced cultures during the same period of time at different place, i.e. Neoanthropines producing neater tools during the same time. Eoliths served as a very good example of some industry ancestral to these Palaeanthropine cultures. But the concept of parallel phyla was long discarded. Thus, the occurrence of Eoliths together with Pleistocene still remains an intriguing problem. The present day archaeologists seem to be in agreement about the incredibility of anything known as Eolithic age.

Life of Eolithic Man

The Eolithic man was probably leading his life in group. His time was spent in food gathering and protection from wild dangerous animals. His food consisted of wild root, leaves, flowers, fruits and small insects. These things were easily available in the environment in which he lived. Probably he used to hunt small birds and animals. His other relatives belonging to primate order were vegetarian. But Eolithic man had started eating meat and had developed non-vegetarian food habit. Australopithecus Africanus, who was related to this age, was non-vegetarian. Non-vegetarian food habit made the human life easy at those places where root and fruits were not available. Now man could live at those places with non-vegetarian food. There is no any evidence to support that whether he knew the use of fire or not.

Fig. 7.1: Eolithic Tools

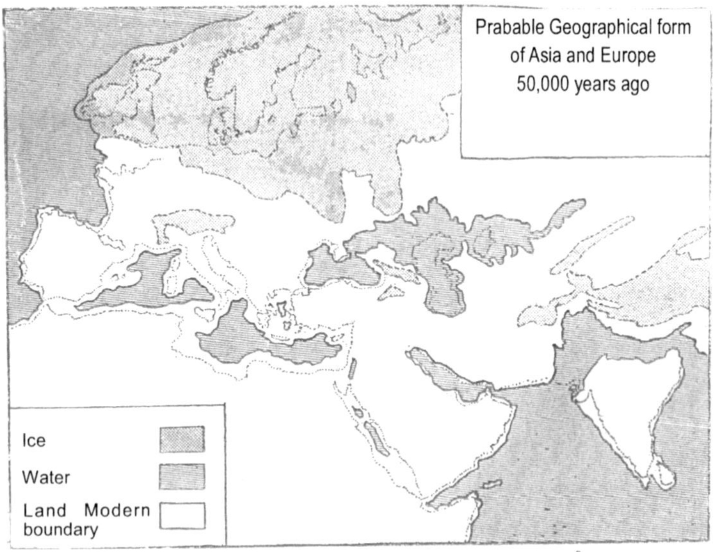

Map 1: Probable Form of Asia and Europe 50,000 Years Before Present

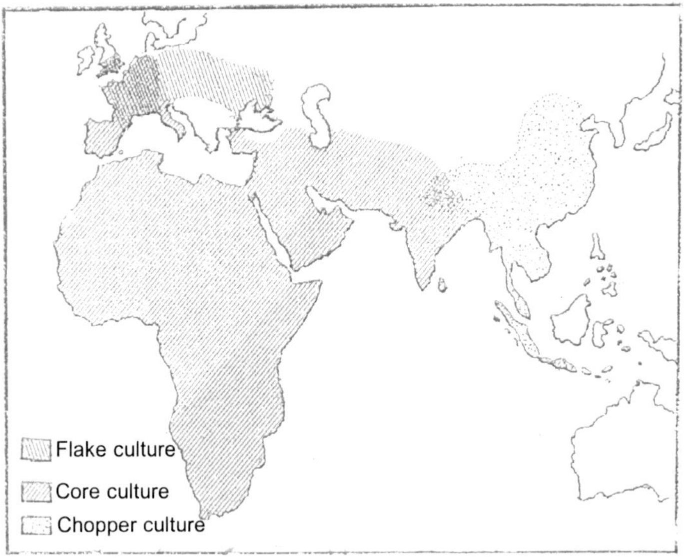

Map 2: Impact Area of Lower Paleolithic Cultures

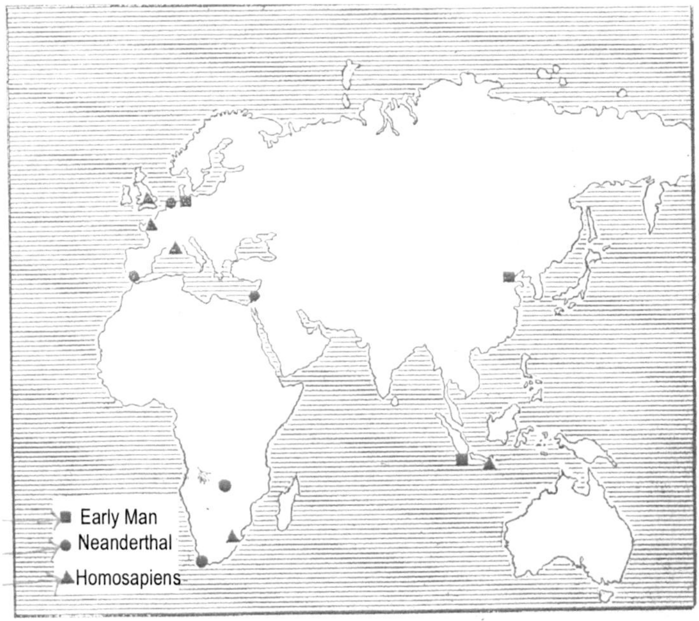

Map 3: Sites of Fossil Remains of Early Man

8

Evolution of Early or Lower Palaeolithic Man and Culture

Early Place of Human Evolution

We have seen earlier that the writers of first chapter of human civilization were not Homosapiens rather they were hominid. They led the foundation of early Palaeolithic and middle Paleolithic cultures nearly 5-6 lakhs to 30-35 thousand years ago in different parts of the world. In fact 95 per cent parts of human history are the history of hominids. But unfortunately we have not been able to know as to where hominids evolved for the first time in the world. On the basis of bone fossils discovered from Peking and Java, Asia was believed as the place of origin of early man. But the discovery of piltdown man suggested Europe as the place of origin of early man. A number of scholars believe today Africa as the place of origin of early man. Thus, there is no unanimity among the scholars regarding the place of origin of early man.

Human Species

Australopithecine

The meaning of the term Australo is south. First of all fossils of human species were discovered from Africa, that is why they have been grouped together under the name Australopithecine. All fossils placed under the name Australopithecine posses some common characteristics. They have been placed in between Apes and man. Prof. Raymond Dart had studied the fossils of Australopithecine for the first time. That is why they are termed as Dartains.

Geographical Distribution

Different fossil finds of Australopithecine have been discovered from Africa. Besides Africa, some teeth and jaw have been discovered from Sangiran in Java. But they were known as Maganthropus.

In South Africa all fossils have been discovered from lime mines. First of all fossils were discovered from Taungs. Then fossils were discovered from Kromdraai, Sterkfontein, Swart krans.

Its fosils have also been discovered from Makapanasgat.

From East Africa fossils have been discovered from Olduwai George. The important members of Australopithecine family are as follows:

	Region		*Name*
1.	South Africa	(a)	Australopithecus Africanus
		(b)	Palesianthropus transvaolensis
		(c)	Paravanthropus Robustus
		(d)	Australopithecus Promethus
		(e)	Paranthropus Crassidens
		(f)	Telanthropus Capensis
2.	East Africa	(a)	Zinjanthropus Boisei
		(b)	Homo Habidlis
		(c)	Australopithecus
3.	West Africa		Australopithecus
4.	Palestine		Australopithecus
5.	China		Australopithecus
6.	Java		Maganthropus Palaeojavanicus

Robinson has classified the members of Australopithecine in two groups which are as follows:

I. Paranthropus Group
II. Australopithecus Group

Paranthropus group consists of the following members:

(a) Paranthropus Robustus
(b) Paranthropus Crassidens
(c) Telanthropus Capensis
(d) Maganthropus Palaeo Javanicus

Australopithecus group consists of the following members:

(a) Palesianthropus Transvaolensis
(b) Australopithecus Promethus
(c) Australopithecus Africanus
(d) Homo Habilis Momo Canomonisis
(e) Homo Canomonisis

Here, we are going to describe three representative members of the family australopithecine namely Paranthropus, Australopithecus Africanus and Homo Habilis.

Paranthropus

The scholars have placed two species in the genus Paranthropus. They are Paranthropus Robustus and Paranthropus Crassidens. Paranthropus Robustus was discovered by Dr. Broom from a place Cromdroai situated nearby Sterkfontein. He named this fossil as Paranthropus Robustus. In the year 1948 evidence of Paranthropus Crassiden was discovered from Sweetkran. Some anthropologists have placed Zinjanthropus Boisei under Paranthropus Robustus. Zinjanthropus was discovered by Leakey in 1959 from Oldowai George of Tanzania in East Africa. The fossil evidences of Paranthropus Robustus include some skulls, humerus, ulna, palm

Fig. 8.1: Paranthropus

bones, etc. On the basis of fossil finds, characteristic features of Paranthropus Robustus are as follows:

- Cheekbones are elevated forward.
- The lower part of upper jaw is flat
- Lower part of face of Paranthropus Robustus resembles with Orange-utan but its Palate part resembles with man.
- Its upper jaw shows Prognathism
- Its lower jaw is heavy.
- The lower jaw of female paranthropus possesses chin.
- The canine teeth of Paranthropus Robustus are similar to man.
- Molar teeth of Robustus are bigger than modern man, but they are full of human features.
- Segital crest was fully developed in the lower jaw of Paranthropus Crassidens.
- Head was not fully developed in frontal bone, but supra orbital torus was fully developed.
- Pelvic girdle of Paranthropus Robustus resembles with man in shape and size, but its ischial bone resembles with apes.
- Paranthropus was able to stand erect and walk with feet.
- Cranial capacity of Paranthropus Crassidens was 1050 C.C. and the cranial capacity of Paranthropus Robustus was 600 C.C. This shows that crassidens species possessed more developed brain than Robustus species. Its period was lower Pleistocene in geological time scale.

Australia Pithecus Africanus

A scholar named Raymond A Dart discovered a child skull from the place Taungs in Rhodesia (Afria). It belonged to lower Pleistocene period. This skull belonged to such an animal which was ape but resembled with man to a great extent. This animal was named as African Ape or Australopithecus Africanus. Again in 1935 a number of fossil evidences of this animal were discovered. The study of those fossil finds indicated that the teeth of that animal resembled with man. According to Dr. Broom, this animal was capable of bipedal walk and probably used some kind of tools of wood sticks and stone pieces. In 1955, Dart obtained some stone pieces from the habitation of Australopithecus. These stone pieces would have been tools of

Fig. 8.2: Australopithecus

of Australopithecus Africanus. Its cranial capacity was 500-700 C.C. According to Sir Arther Keith, Australopithecus Africanus was the early ancestor of modern man. Bones discovered nearby Johansberg supported this view of Arther Keith. Majority of bones belonged to Australopithecus. One bone, though resembled with Australopithecus Africanus, but was superior to it. This animal was named as Talenthropus. This bone became helpful in establishing Africa as the place of origin of man. But majority of scholars do not accept Australopithecus Africanas as the ancestor of man. The cranial capacity of Australopithecus was very small in comparison to man. The structure of skull of Australopithecus resembles with Gorila to a great extent. It seems probable that Australopithecus was not early ancestor of man, rather he was closely related to that ancestor of man. Its geological period was lower Pleistocene. The chief characteristics of Australopithecus Africanus are as follows:

- Supra orbital ridges were absent in the skull of Australopithecus Africanus.
- Circular eye orbits resembles with Orang-utan and Lemurs.
- Head was dolicocephalic.
- Nasal aperture was small and equal in length and breadth. This characteristic was very much similar to man.
- Jaws showed prognathism.

- Teeth were fixed like man.
- Canine teeth were small and tusk like which resembled with apes.
- Diastema was present in between canine teeth of upper jaw.
- Palate was oval in shape.
- Lower jaw was solid and thick like man.
- Chin was absent.
- Foramen magnum was located forwardly in comparison to apes.
- Its head was more balanced than apes.
- The cranial capacity was 520 CC.
- Its head was bigger than Chimpanzee and resembled with man.

Evolutionary Status: From the above description, it appears that Australopithecus possessed some characteristics of ape and some characteristics of man. But majority of characteristics resemble with Homosapiens. He possessed bipedal walk and used tools. In this way this hominid was closely related to man. There are three main opinions related to Australopithecus. These are:

1. Australopithecus Africanus was like apes. Its physical structures resembled with Chimpanzee and Gorilla, though its certain characteristics resembled with man.
2. Although some physical features of Australopithecus Africanus resembled with man, but it was not our early ancestor. It was closely related to our ancestor.
3. Australopithecus Africanus occupies its evolutionary position between apes and man.

Homo Habilis

The fossil evidence of Homo habilis was discovered by Leakey in 1962 from Oldowai George situated in East Africa. Its fossil evidences have also been discovered from Kubifora place in Kenya. It was named as Homo Habilis or hand using man. The meaning of the term Habilis is Handman. In this way this animal was related to tool making. Its geographical distribution was limited only to East Africa. The fossil evidences of Homo Habilis include some parts of skull, lower jaw, bones of feet, some bones of hands, clavicle, etc. On the basis of those evidences, Homo Habilis possessed the following physical characteristics:

- Its bones of hands resembled with Gorilla.
- The structure of bones of feet reveals that Homo Habilis was a bipedal animal.
- Homo Habilis is more closer to man than apes.
- The structure of bones of hands indicates that Homo Habilis possessed strong holding efficiency. It used to spend night on trees and during day time they lived on land.
- Its skull was somewhat high and round in shape.
- Its face was flat and less elevated outside.
- Its molar teeth were big in shape and covered with thick layer of enamel.
- Its premolar teeth were small in size.
- Its grinded premolar and molar teeth bear testimony that Homo Habilis was a vegetarian animal.
- Its cravial capacity was 660 CC to 775 CC.
- It was very close to man physically.
- It was the first cultural animal of the world.
- Its geological period was lower Pleistocene.

Evolutionary Status: According to scholars working on the fossil evidences of Homo Habilis, Homo habilis occupies position in between Australopithecus and Homo erectus. But some scholars do not agree with this opinion. They consider Homo Habilis as a special type of Australopithecus. They are not quite different from Australopithecus. Its evolutionary status can be considered as ancestor of Homo erectus.

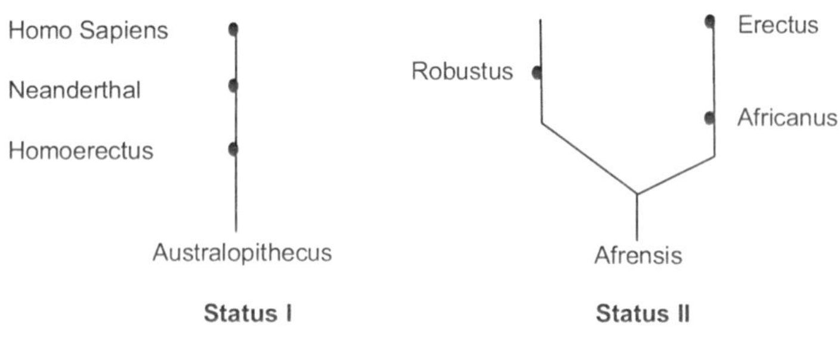

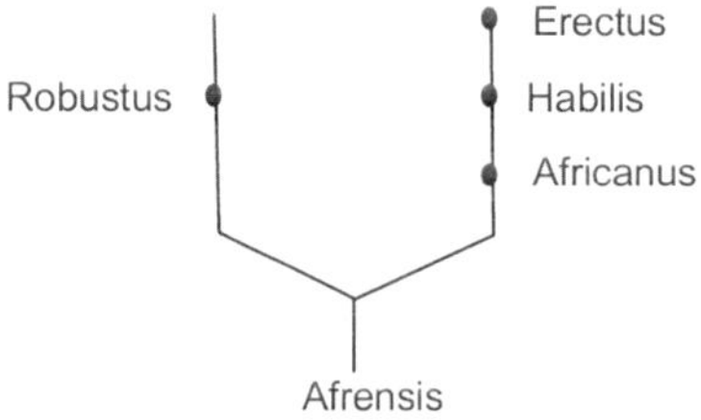

Status III

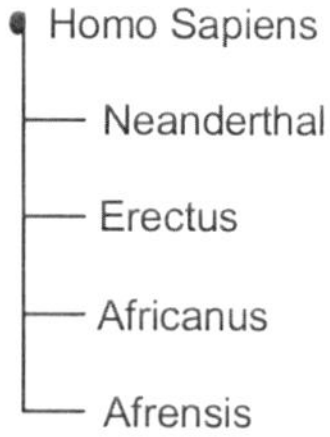

Status IV

Homo Erectus

The Hominids of Asia are called Homo erectus. As discovery of fossil evidence of Australopithecus from Africa is taken as the early place of human evolution, in the same way discovery of fossil evidences of Homo erectus from Asia is taken as the place of evolution of early man.

Fig. 8.3: Homo Erectus

The credit for discovering fossil remains of early man in Asia goes to a scholar named Eugene Dubois. In 1891, he reported about the fossil finds form a place Trinil of Java on the bank of river Solo. From this place he was successful in discovering a thigh bone, a skull and two teeth as fossils. The man of these fossilized bones was named as Pithecanthropus erectus. The cranial capacity of this man was 900 CC. This man is also known as Java man. It possessed erect posture and bipedal walk. Its cranial capacity was more than Chimpanzee and less than Homo sapiens. It was certainly more developed than known Anthropoid apes. Its cranial capacity was even more than Australopithecus. He could stood on feet, he could run and could move his hands independently. Probably this man knew to speak too.

In 1936, a human child skull was discovered from eastern Java. Its geological layer belonged to earlier than Pithecanthropus erectus. The man with these bones was named as Pithecanthropus Myjokertensis. The child skull of this man was more developed than anthropoid apes. The time span of these two Java men has been established as between 8 lakh to 5 lakh years ago.

Contemporary to Java man or even earlier than him, human fossils have been discovered from a place Choukoutein in China. This place was situated at a distance of 37 miles from Peking. 40 skeletal remains including 40 skulls were discovered by a Chinese scientist named W.C. Pei. The man with those bone was named as Sianthropus Pekinensis. It was also known as Peking man. This man like Java man possessed erect posture and bipedal walk. Peking man resembled with Java man in several characteristics. But its cranial capacity was greater than Java man. Its cranial capacity was 1075 CC. Its speech region was also more developed than Java man. Along with the bone remains of Peking man, several bone remains of other animals have also been found. Evidences of use of fire have also been discovered.

Before 1960, Scientists believed that Java man and Peking man represented two species. But now they believe that both belong to the same species, but represent two geographical form. Both belong to the same genus Homo. That is why they have been named as Homo erectus. Discoveries made after World War Second revealed that they had spread upto Asia, Africa and Europe.

Pithecanthropus Erectus (Java Man)

As stated earlier, first discovery of pithecanthropus was made by Professor Eugene Dubois in 1891. According to geological time scale, it belonged to lower Pleistocene period. Its chief physical features were as follows:

- The length and breadth of the skull of Java man was found 18.5 cm and 13.0 cm respectively. Its celephalic index was measured as 70. It was placed under Doliocephatic group.
- Like apes, its skull was flat frontally and vault was situated downwardly.
- Nuchal muscles on occipital bone present a situation between ape and man.
- Like apes, supra orbital torus was present.
- Foramen magnum was more forwardly located than apes.
- Median keel was present on frontal bone.
- Temporal lines parallel to vault were found.
- Like apes brain was less developed.
- Teeth were greater than man.
- The roots of molar teeth were strong and divided into two parts.
- Crown of teeth resembled with man.
- The length of femur bone was 45.5 cm. On the basis of length of femur bone, the body length of Java man has been estimated between 167 cm to 170 cm.
- Cranial capacity was found as on 940 C.C. This indicates a situation between ape and man.
- Linea aspera of femur was well developed. This indicates clearly that Java man was capable of standing erect and bipedal walk.

Evolutionary Status: There is no unanimity among the scholars regarding the evolutionary status of Java man. According to one view, Java man occupied evolutionary status between ape and man. According to another view, Java man was ancestor of man in Pleistocene period who gave rise to Neanderthal man.

Sianthropus Pekinensis (Peking Man or Chinese man)

In 1926, fossils of two teeth were discovered. In 1927 some other teeth were also found. They were studied in detail by Professor Davidson Black. The man with those fossils was named as Sianthropus

Pekinensis. In 1929, Dr. W.C. Pei discovered a full skull of this man which was named as Sianthropus skull no. 1. This man belonged to lower Pleistocene period in geological time scale. The main physical features of this man were as follows:

- The average length of skull of this man was measured as 19.4 cm. General head length was measued between 16.5 cm to 20.5 cm.
- Head breath of Chinese man was measured between 13.7 cm to 14.3 cm.
- Cranial length was found as 72.2. The head belonged to dolico cephalic group.
- Brain vault was measued as 11.5 cm which was comparatively very long.
- The cranial capacity of Chinese man was observed in between 850 CC to 1300 CC. The average cranial capacity was reported as 1075 CC.
- The skull bones of Chinese man were more thick than apes.
- The sutures of skull resembled with ape.
- Supra orbital ridges were continuous.
- Medial keel was present.
- Eye orbits were big.
- Nose was more flat than modern man.
- Nasal spine was not present.
- Molar teeth were big and specially resembled with Mongoloid people.
- Palatal arch was narrow in comparison to man and wide in comparison to ape.
- Face was flat and small.
- Jaw of male Sianthropus was big in comparison to modern man.
- The jaw of female Sianthropus Pekinensis resembled with Mongoloid people.
- Chin was absent.

Evolutionary Status: There are two opinions regarding the evolutionary status of Chinese man. According to one view Chinese man stood evolutionally between ape and man. But according to an another view Chinese man was ancestor of modern man in Pleistocene period who gave rise to Neanderthal man.

Hominid of Europe: Till 1952 some scientists held view that like Africa and Asia, Europe, too could be considered as the early place of origin of man. Main base behind this belief was discovery of human fossils from piltdown from Sussex region of England. Its geological period was lower Pleistocene. This human fossil was reported by Charles Dowson in 1912. On the basis of study, it was found that the skull and teeth of this animal resembled with man but its jaw resembled with ape. Evolutionally, it was placed in between ape and man. Anthropologists of England termed it proudly as missing link of human evolution. It was named as Eoanthropus Dowson. It was also known as Piltdown man. But scholars of America and Europe were not ready to accept it as missing link of human evolution because skull and jaw were unsuitable in look. With the passage of time, this doubt turned in to belief. Its jaw resembled with ape and skull resembled with man. In 1952, fluorine test of skull and jaw remains was done. This test revealed the fact that piltdown man was not a reality. Rather it was a great scientific hoax. Its skull was not of 5 lakh years old, but it was merely 50 thousand years old. Its jaw was comparatively newer.

Heidelberg Man

When secret about the piltdown man was opened, one does not find any other human remains of ancient man in Europe contemporary to Java man or Chinese man. Therefore, at the present level of knowledge, one cannot consider Europe as the early place of origin of man.

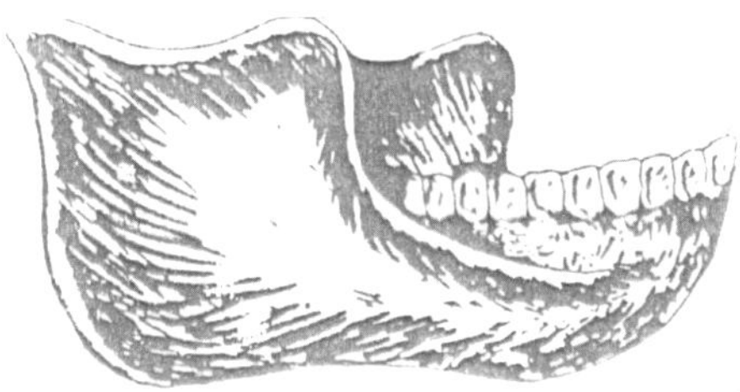

Fig. 8.4: Heidelberg Jaw

The earliest human remains from Europe available to us is Heidelberg man. It belonged to early Pleistocene period. The jaw of this man was discovered by Dr. Otto Schoetensack in 1907 from a place Mauer 6 miles away from Heidelberg in Germany. It was discovered from a mine named Mauer. The jaw belonged to second glacial period or beginning of IInd Interglacial period. Heidelberg man resembled with Pithecanthropus erectus. But it was more developed than Pithecanthropus and was nearer to modern man. It is also known as Homo Heidelbergensis. Probably it gave rise to Abbevillian culture. With Heidelberg jaw, tools of lower Paleolithic culture were also discovered. The chief physical features of Heidelberg man were as follows:

- Its jaw was big, thick and strong.
- Its dentition was like man.
- Its jaw was oval in shape.
- Chin was not present.
- It resembled with man and apes.

Evolutionary Status: Its evolutionary status is doubtful. But it could be considered as ancestor of Neanderthal man.

Early Homosapiens of Europe: After Heidelberg man, one finds the emergence of age of those man whose remains were discovered from Swanscombe, Steinheim and Fontechevade. On the basis of places from where human remains have been discovered, they were known as Swanscombe man, Steinheim man and Fontechevade man. They are also termed as early Homo sapiens.

Swanscombe Man

There is a small town in Cant region towards south of Tames river. This place is known as Swanscombe. In 1935 a scientist named Mastern had discovered a human skull from the geological layer of second interglacial age. The man with remains of skull was termed as Swanscombe man. On the basis of study, it was found that the skull belonged to a woman whose age would have been 20 years at the time of death. Her cranial capacity would have been 1325 CC to 1350 CC. Her cranial capacity resembled with the cranial capacity of modern woman. This woman was placed under Homo sapiens. She belonged to middle Pleistocene period. She resembled with English woman.

Steinheim Man

The skull remains of Steinheim man was discovered from Stuart district of South Germany. It belonged to later phase of second interglacial period. Its geological period was middle Pleistocene. Probably this skull belonged to a woman. Cranial capacity of this man was 1100 CC. But in other characteristics this man could be considered as Homo sapiens.

Swanscombe man and Steinheim man did not represent Neanderthal man.

Fontechevade Man

In 1947 Kumare Henary Merton discovered two human skull remains from a cave named Fonntechevade in South France. These remains were found at a depth of 10 feet. Their geological period was middle Pleistocene. These skull possessed very much similarity with modern man. In thickness, they were different from Neanderthal man. But they resembled very much with Swanscombe man. The maximum width of the skull like Swanscombe man was 154 mm. The skull belonged to mesocephalic group. Cranial capacity was 1450 CC. Forehead of Fontechevade man was very much like to modern man.

Evolutionary Status: Swanscombe man, Steinheim man and Fontechevade man belonged to period earlier than middle Palaeolithic period which is known for the emergence of Neanderthalman. These men had emerged earlier than Neanderthal man. Thus, they were older than Neanderthal man. But physically they were very much similar to Homo sapiens or modern man. They occupy evolutional status as early Homo sapiens.

Tools of Early Palaeolithic Period: From the last part of early Pleistocene period or from the period of first inter glaciation, one finds the emergence of such stone tools about which one cannot doubt that they were not made by man. Among these tools coup-de-poing or hand axe occupies the earliest place. This tool was pointed from the front side with sharp edges in lateral sides. The back side was round so that it could be held in hand easily. In the early period, this early tool was only means of man by which he took the work of hammer, scrapper, knife, axe, awl, etc. This early stone tool was helpful in hunting animals, scraping skin and digging out root. But as the experience of man went on increasing, he made different types of tools for different kinds of activities. These tools may be classified into three categories:

1. Core Tool.
2. Flake Tool.
3. Choper Tool.

Core Tool

In order to make core tool, a stone fragment was flaked in such way that its middle part could be used as a tool. The middle part of the flaked stone fragment is called core. As this was used as tool, so it was named as core tool. Thus, core is a starting material of stone from which flakes are removed. When it is given a shape of tool by flaking, it is known as Core Tool. Core tools are characteristics of early Palaeolithic period. They have been discovered from Africa, Cyria Palestine, Western Europe (Spain, France and England) and South India.

From the point of view of evolution, the Early Paleolithic core tools are divided into three cultures, namely, Early Chellean culture, Chellean or Abbevillian culture and Acheulian culture. As these cultures were discovered and studied for the first time in France, therefore, they have been named after those places of France.

1. *Early Chellean Culture*: This name has been given to this culture on the basis of tools discovered from a place situated at a distance of 8 miles from Paris in France. But this cannot be accepted that the place of origin of this culture is France. In fact this culture originated for the first time in middle Africa. Gradually it spread to Western Europe and South Asia. The coup-de-poing or hand axe tools of this culture are quite simple and heavy. Sometimes, they resembled with eoliths or natural stone fragments. Because no refinement has been shown in their making. These tools belonged to the period of second glaciation. Probably Pithecanthropus man was wandering on the earth in this period.

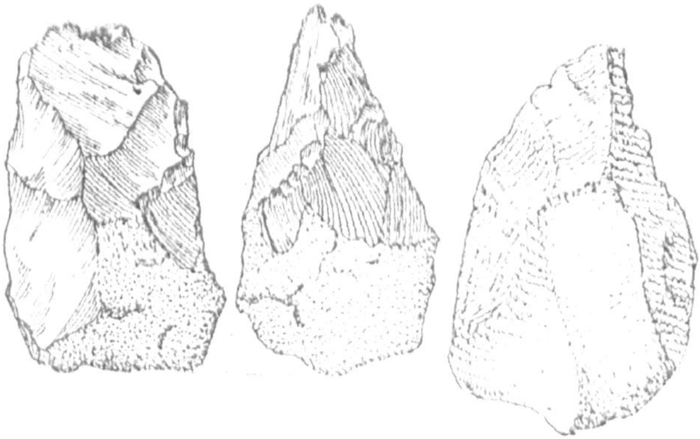

Fig. 8.5: Chellean Hand Axes

2. *Chellean or Abbevillian*: After the period of early Chellean culture, one finds the beginning of Chellean or Abbevillian culture. The time span of this culture has been established as till the beginning of second interglaciation. In this period, hand axes of early Chellean period were flaked from both sides so that they could be more useful. During this period, the descendants of Peithecanthropus (Java man) or Heidelberg man were wandering. Chellean hand axes have been discovered from England, France, Spain, Portugal, Palestine, India, Java and Africa.
3. *Acheulian Culture*: This term takes its origin from St. Acheul, a suburb of amien in France. The time span of this culture has been established as from middle of second interglaication to third interglaciation. The tools of this culture were more pointed and beautiful than the tools of earlier cultures. The hand axes of this culture had become pear shaped. Acheulian men knew to make tools from the flakes which they attained at the time of making core tools. But the main tools of Acheulian man were hand axes. The tools of this culture have been discovered from Europe, Greenland, America, Canada, Mexico, Western Asia, India and China. In this cultural periods, those men were surviving whose remains have been discovered from Swanscombe, Steinheim and Fontechevade. i.e., Early Homosapiens.

Fig. 8.6: Acheulian Culture

Flake Tool

Flake represents a part of the core with a flake surface with or without bulb of percussion and a platform. Based on the location of striking platform, flakes are called side flakes, end flakes and intermediate flakes. The tools prepared from flakes are called Flake Tools. In making a flake tools, core is left. Only flakes removed from the core are used. There are many types of flake tools. They have been discovered from Europe and Northern Euresia. Since flakes were removed from cores, it is quite natural that flake tools were made along with core tools since early time. On the basis of evolution, flake tools have been classified into following cultures:

- Clactonian Culture
- Levalloisian Culture

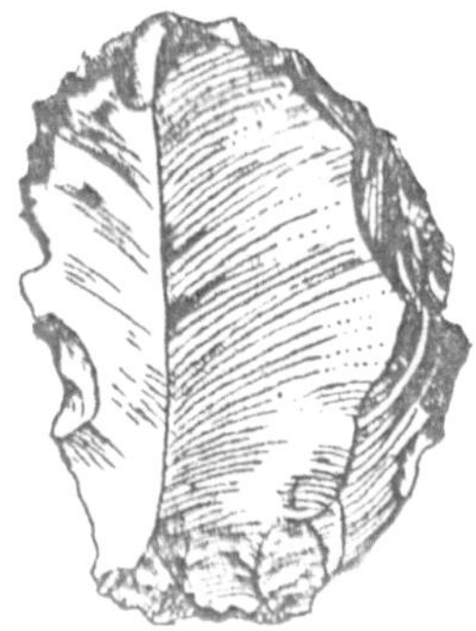

Fig. 8.7: Clactonian Flake

Fig. 8.8: Levalloisian Culture

Clactonian Culture: It is named after a site Clacton on sea. In Europe, the history of this culture begins from second glacial age. It continues after Acheulian culture in second interglacial period. Although one finds tools of Clactonian culture at a place Swanscombe, but when men of Acheulian culture settled in Europe, one finds the meeting of both these cultures there. In Middle Europe, the impact of Clactonian culture remained intact.

Levalloisian Culture: It is named after a locality at Levallois, a suburb of Paris. For making tools of this culture, the shape of tool is engraved in the stone fragment. Then it is separated by breaking. This technique of tool making developed in different parts of the world independently. The period of this culture is from the beginning of third glacial age to third interglacial age.

Chopper Tool

When core and flake cultures were flourishing in Europe, Africa and some parts of Asia, at the same time North-Western India, South-East and East Asia were witnessing the emergence of an another culture known as Chopper culture. Although Chopper Tools resembled with hand axes, but at the same time they were slightly different. In this edges were made only in one side. Majority of choppers were made up of flakes, but sometimes cores were also used for this purpose. Chopper tools have been found from Patajitanian culture of Java, Soan culture of North-Western India, Annathian culture of

Burma and Cloukouteinian culture of China. These tools have also been found in Stellenboselh culture, Early Oldowan culture and Kafuan culture of Pre-chelleanage Africa.

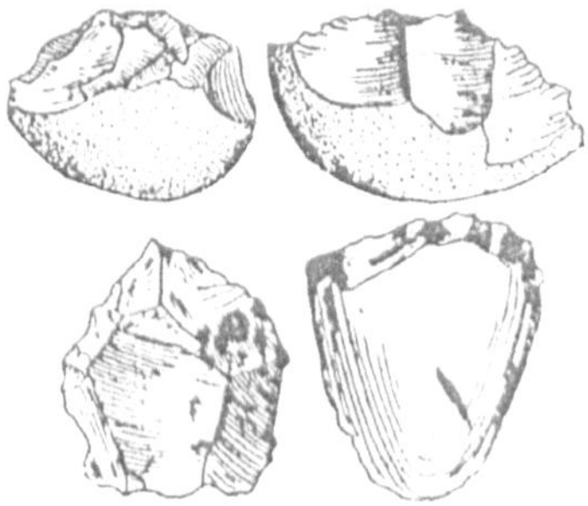

Fig. 8.9: Chopper Tools

From the above description, it is quite clear that during lower Palaeolithic age three cultural currents were flowing in parallel manner. In Chellean-Acheulian cultural tradition, core tools occupied prominent place. In Clactonian-Levalloisian culture tradition, flake tools occupied significant place. But in Soan-Annathian, Patjitanian-Choukouteinian cultural tradition, chopper tools occupied important place.

Flake tools were more useful in glacial environment. Core tools were more useful in interglacial environment. That is why in Europe flake tools were more popular in glacial periods and core tools were more popular in interglacial periods.

Social-Cultural Scenario of Australopithecus

Some tools have also been discovered from bone remains of Australopithecus. These tools were found from Swarkrans, Sterkfontein, Oldowan (Fig. 8.10) and Amono valley along with animal fossils. The tools were big ugly and without any shape and size. Among these tools pebble chopper tools were more in number. These tools were made by flaking a general stone fragment or pebble from one side or both sides. From Oldowan eleven types of such tools were discovered. These tools were used for cutting, scraping and digging. One bone tool has also been discovered from this place. It seems probable that Australopithecus used bone tools made up of bone, horn, teeth and jaw.

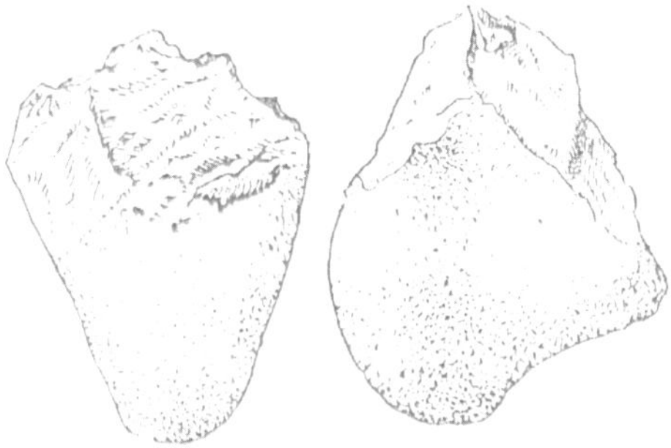

Fig. 8.10: Oldowan

Bone remains of small animals have also been discovered from Oldowan. These fossils have been dated to 17 lakh years. These bones would have been used for taking out marrow during early Pleistocene period, the African regions were inhabited by numerous animals. For example deer, elephant, rhinoceros, yak, horse, ziraffe, bear, lion tiger, etc.

Meat of animals was important food items of Australopithecus. They used to hunt small animals with the stone tools they had made. Probably they had developed the concept of cooperation for the purpose of hunting. The concept of cooperation would have given the rise of early social organisation. Probably symbols and language would have emerged for transaction of reciprocal activities.

The habitat of Australopithecus was situated below sky. These used to wander around lakes and rivers. They had not started living in caves. The excavation of Oldowan George has revealed the evidence of place of residence of Australopithecus. This habitat occupied an area of 3400 sq. ft. More than 400 bone fossils and stone tools have been discovered from this site. A circular habitat was also found. This acted as resistance against wind. These evidences reveal a definite type of habitat of Australopithecus. But nothing can be said with certainty about their stability.

The earliest evidence of use of fire in Africa belonged to Acheulian culture period.

Some assumptions regarding the social life of Australopithecus have been formulated. According to these assumptions Australopithecus lived in group. The number of members in this group would not have been more than 25. They practised infanticide. 15 per cent to 50 per cent children were killed in order to control the population of the group. Division of labour on the basis of sex would have been prevalent. Males remained engaged in hunting. Food collection and rearing of children were done by females. The family would have been female centred. The average age of Australopithecus would have been 18 to 23 years.

Social-Cultural Scenario of Java Man (Pithecaunthropous erectus) and Chinee man (Sianthropus Pekinensis)

Several remains of bones and tools have been found associated with Java man and Chinese man. These evidences throw light on the social-cultural life of Java man and Peking man. They were cave dwellers. They had survived for a long period of time. Evidences of burnt bones reveal that Java man and Chinese man were aware of use of fire. The fire was used as a tool to protect from cold and dangerous wild animals. Fire was also used for roasting the meat of animals hunted.

The base part of some skulls of Java man and Chinese man were found broken. This reveals evidences that they were man eater. It is also probable that those part of skull would have been broken for the purpose of some rituals they used to perform. From some places remains of chest bones and vertebral column have been discovered along with skull remains. These acts would not have been for the fulfilment of needs but for the performance of rituals. They did not eat meat of man.

The tools associated with Java and Chinese man were pebble choppers, flake tools and beaked scrappers. These tools were made by soft hammer. They resembled pear or they were pearshaped. They are termed as core tools. The long, fine and shaped tools were used for the purpose of cutting, scrapping, digging and boring.

The life of Java man and Chinese man was fully dependent on hunting and gathering. Bone remains of big animals like rhinoceros, deer, bear, horse, lion, tiger, etc. were found along with the tools of Java man and Peking man. These bone evidences suggest that they were practising hunting of big animal. They used to dig out long pit to trap these big animals. They used to kill those animals with tools they had made. After killing they used to eat meat roasted in fire. They used to scrap skin and cut meat into pieces.

Such type of hunting would have required organised activities and living in group. They would have been living in group. They would have been doing hunting in organised group. Because it was very difficult at that time to hunt big animals without any kind of organisation.

There would have also been division of labour based on sex. Female would have been kept engaged in rearing infants and doing collection of food materials. Males of the group would have been engaged in group hunting.

The living in group would have compelled them to develop symbols and language for communication and expression of things, events and activities.

Culture would have become an integral part of Java and Chinese man. They would have become a cultural man.

Social Cultural Life of Early Homosapiens

Swanscombe man, Steinheim man and Fontechevade man are known as early Homosapiens. They resembled with Homosapiens, Sapiens more than Neanderthal man, but existed earlier than Neanderthal man in lower Paeleolithic period.

Evidences suggests that they had evolved organisation and lived in group. They used to live in cave. They subsisted on hunting and gathering. They used to prepare chopper tools along with some flake and core tools. They used to posses some kind of symbols and language for activities, need fulfilment, communication and expression. Division of labour based on sex would have emerged in them.

9

Evolution of Middle Palaeolithic Man and Culture

Human Species: Neanderthal Man

Middle Palaeolithic human species have been termed as Neanderthal man by the scientists working in Palaeo-anthropology. This man used to inhabit in Europe, Northern Africa, Eastern Africa, and Asia. In Europe, the man was living in Belgium, France, Spain, Italy, Yugoslavia, Crimea etc. The geological period in which Neanderthal man lived was from the last part of third interglaciation and till middle of Wurm (Fourth glaciation) period. Thus, Neanderthal man would have lived from 100,000 years to 35,000 years ago. They were found living in cold regions. They used to live in cave. Majority of remains of this man were discovered from cave sites. They had spread up to Crimea, Lebanan; Palestine and Iraq. They occupied position in between Australopithecus and Homo erectus.

Fig. 9.1: Neanderthal Man

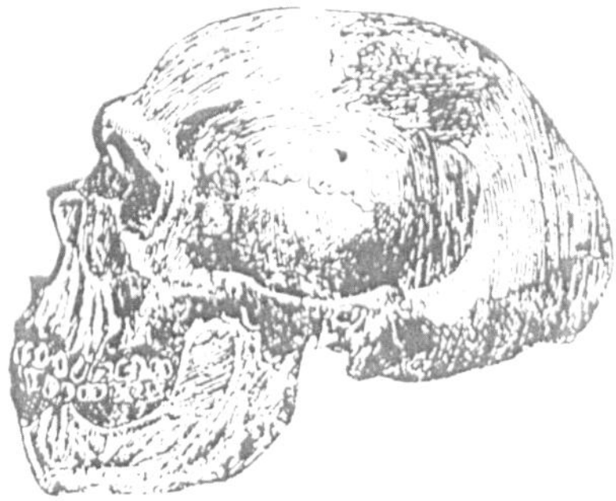

Fig. 9.2: Neanderthal Skull

Discovery

Fossil remains of Neanderthal man was discovered for the first time in 1848 beneath a rock in Gibraltar. But no one could pay academic attention towards this fossil remain. After that in 1856 some skeleton remains were found from the place Neanderthal in Dusseldorf region of Germany. On the basis of this place, the man was named as Neanderthal man.

Again after the discovery made at Neanderthal, skeleton remains of man belonging to the same period were discovered from Belgium, Spain, France, Italy, Yugoslavia and Crimea. The man found from these places were placed under the species Neanderthal man. On the basis of similarity in physical features, all remains of Neanderthal man have been classified into two categories which are as follows:

- Conservative Neanderthal
- Progressive Neanderthal

Conservative Neanderthal: This type includes several remains discovered from La-Chappele-aux-Saints, La-Moustier, La-Quena, La-Ferrasie etc. Here the description of La-Chapelle-aus-saints is being as representative of conservative Neanderthal type. Its importants features are as follows:

- Its cranial capacity was measured as 1600 CC.
- Its skull was big and heavy.
- Its head length and breath were 208 mm and 155 mm respectively.

- The average cephalic index was 74.5. It varied individually from 70 to 76.
- Face was more developed in comparison to head.
- Frontal part of head was comparatively big.
- Vault height was low.
- Its head was less developed.
- Occipital region was found elevated and depressed.
- Temporal fossa was found big.
- Temporal line spread up to half distance in brain cage.
- Squamosal suture was not very much spiral.
- Continuous and big supra orbital ridge.
- Very big eye cortex was found.
- Torus formation was very much like to Chimpanzee, Gorilla and Peithecanthropus erectus.
- The position of maxillary was strong and heavy.
- Nose was platyrrhine type.
- Nasal root was found depressed like Australians.
- Round lines were found on occipital region.
- Nasal aperture was long.
- Palate was found.
- Mastoid process was very small.
- Glenoid cavity was big and post-glenoid epiphysis was more developed.
- Lower jaw was big and strong.
- Chin was less developed.
- Symphysis slopes were found but simian self was absent.
- Diastema was absent in teeth
- Humerus was strong and small. Its big head resembled with man.
- Hand bones were strong and small.
- Head of humerus resembled with man.
- Hands were smaller than feet.
- Its back bone was small and heavy.
- Clavicle bone was round like Chimpanzee.
- Femur was strong and heavy. Its head was also big.
- Linea aspera like apes was less developed in femur.
- Tibia was small and strong.
- The length of its entire body was 5 feet. But its height was observed between 5'1" to 5'5".

- Its palm resembles more with apes.

Progressive Neanderthal: Remains of Neanderthal man found from Palestine, Yugoslavia and Steinheim, Spain, Belgium, Germany, France, etc. have been placed under Progressive Neanderthal type. They resembled very much with Homo Sapiens or Modern Man.

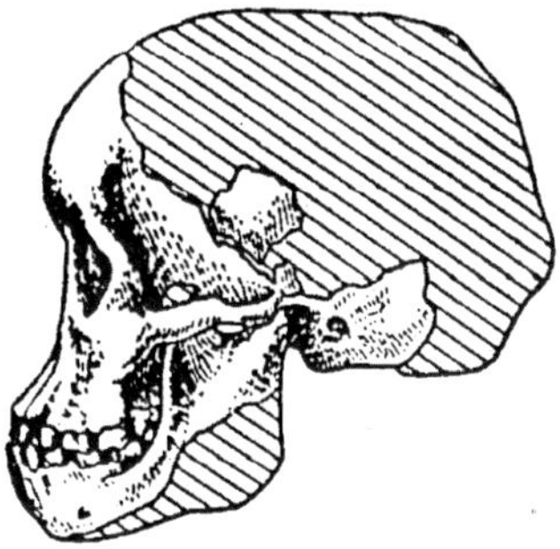

Fig. 9.3: Australopithecus Africanus Skull

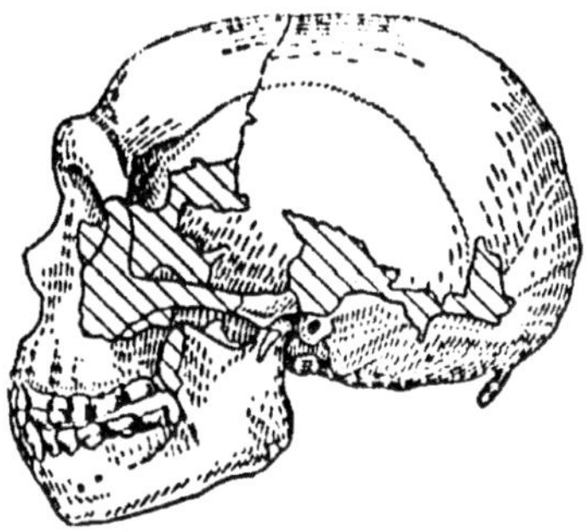

Fig. 9.4: Neanderthal Skull

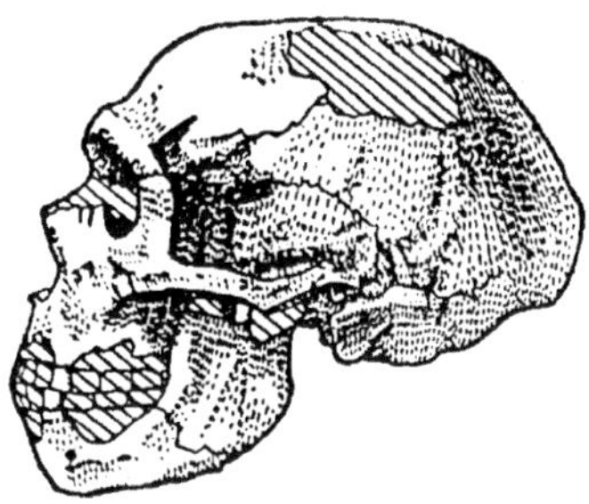

Fig. 9.5: Neanderthal Skull Found from Mount Carmel

In 1931-32 some remains of Neanderthal man were discovered from two caves of Carmel mountain in Palestine. From one cave, skeleton of a young girl along with mandible and other bones were found. From another cave broken fossil bones of ten individuals were discovered. Along with these bone remains some animal remains were also found. In 1899, fossil bones of 20 individuals were discovered from Krapina in Yugoslavia. The skulls were of brachycephalic groups. In 1933, remains of jaw were discovered from Steinheim in Germany. It was related to Acheulian culture. Its geological period was middle Pleistocene. Its head was long and narrow and cephalic index was 70. Cranial capacity was 1070 CC. Although its head was smaller than conservative Neanderthal, but super a orbital torus resembled with conservative Neanderthal man.

Here description of Neanderthal man found from caves of Carmel mountain is being presented as representative of progressive Neanderthal man.

The chief physical features of progressive Neanderthal man are as follows:

- The males of progressive Neanderthal man possessed more height than females.
- The head of progressive Neanderthal was heavy and height of vault was medium.
- Zygomatic process was like modern man.
- Canine fossa was absent.
- Eye orbit was flat.
- Its nose shows speciality.
- Chin was fully developed in some members.
- Upper branch of mandible was less flat.
- Teeth were comparatively smaller.
- Molar teeth show some primitive character.
- Cranial capacity of individuals varied between 1418 CC to 1857 CC in males.
- The cranial capacity of females varied between 1300 CC to 1350 CC.
- The brain of progressive Neanderthal was more developed.

Evolutional Status: The physical structure of Neanderthal man resembles with Homo sapiens (True man or modern man). But one finds some differences in physical features, too. The stature of Neanderthal man was small. Its height varied between 5 ft to 5'5". Head was big, nose flat but pointed, shoulder flat and head sloping downwards. Fingers were not flexible like modern man. He could not stand keeping his neck erect. He could not walk continuously. Cranial capacity was more than modern man. But head belonged to lower category. His brain possessed weak power of seeing and touching. Probably he was able to speak but had not developed language. Although the scholars like Ashley and Montague have attempted to show that Neanderthal man resembled with modern man to a great extent, but some other scholars do not agree with this opinion. They believe that Neanderthal man possessed physical demerits and was not similar physically to man.

On the basis of above description, it is not easy to decide the evolutional status of Neanderthal man. Previously scholars held view that Neanderthal belonged to genus Homo but they were not true man or Homo sapiens. The believer of this view had isolated Neanderthal man from Homo sapiens. According to them Neanderthal was a semi human species. These semi human species were defeated by Homo sapiens of upper Palaeolithic period. The Homo sapiens of upper Palaeolithic period had defeated them and established their control over Europe. But recent discoveries made at Swansecombe, Steinheim and Fontechevade have revealed the remains of such human species who belonged undoubtedly to Homo sapiens. Now we cannot say with certainty that Homo sapiens emerged in Europe as destructor of Neanderthal man. Now it seems probable that in the early period of glaciations such human species came to settle down in Europe who resembled with Homo erectus. The evidence of Heidelberg Man bears testimony in this regard. Those human species gave rise to Homo sapiens in lower Palaeolithic period. But when Europe met with fourth terrible glaciation in middle Palaeolithic period, a branch of Homo sapiens were left isolated. This branch of Homo sapiens represented Neanderthal. Due to isolation in glacial period, some changes in physical features of their body took place. They began to appear different from Homo sapiens. Thus, Neanderthal man was basically related to Homo sapiens.

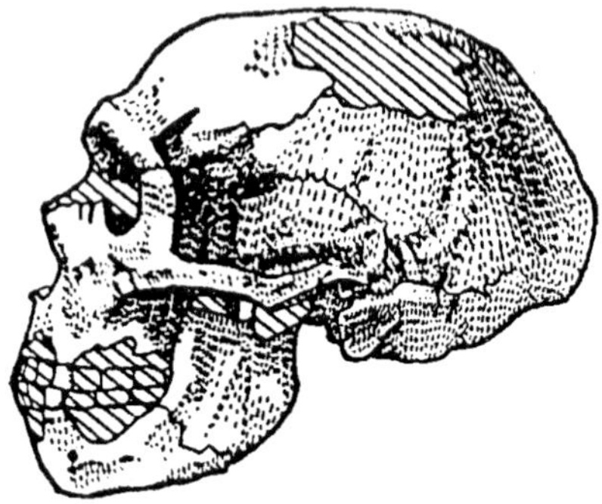

Fig. 9.6: Cromagnan Man Skull

Fig. 9.7: Neanderthal Man

Tool: The tools made by Neanderthal man belonged to Mousterian culture. These tools were discovered abundantly from La-Mousteir, a place in France. They have been termed as Mousterian. Besides France, Mousterian tools have been discovered from several European countries, West Asia and Northern Africa. All these tools are basically flake tools. Although Mousterian tools were developed from Clactonian tradition, but they were also influenced by Levalloisian tradition. They were lighter, finer and beautiful than earlier flake tools. They were result of experiences of several centuries. Among these tools, side scrapper, knife, awl, spear head, etc. occupied prominent place. Neanderthal man used naturally broken pieces of bones as tools. They did not know to prepare bone tools by their activities of hand.

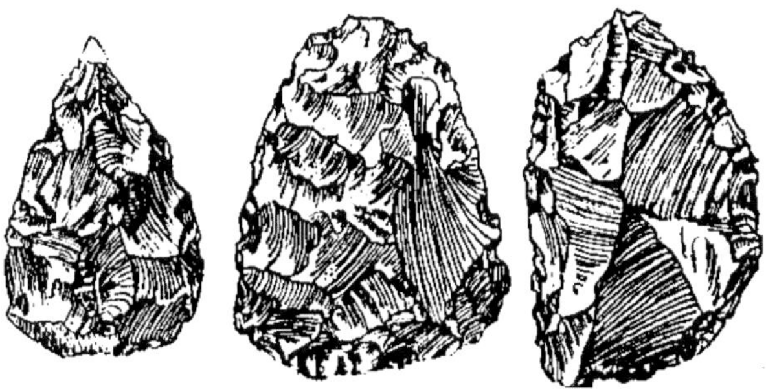

Fig. 9.8: Mousterian Tools

Period of Neanderthal Culture: The period of Neanderthal culture begins from end part of third interglaciation period and continues till middle of fourth glaciation period. In the last part of third interglaciation period, the environment of Europe was hot. That is why the life of Neanderthal of that period shows similarity with Acheulian. But during fourth glaciation period, the life of Neanderthal man shows complete change. The middle Palaeolithic represented a single cultural tradition called as Mousterian. It was defined as culture associated with Neanderthal man (Homo Neanderthalensis). The culture name was taken from a rock shelter in South-Western France called La-Moustier. According to this early definition Mousterian culture was characterised by the presence of points and side scrappers and rarely by hand axes that were thiner than preceding Acheulian stage. This culture was taken for a long time as a simple unilineal development with preceding stage. But with more and more evidences the earlier views seem to be hardly tenable at present. Now we have four Mousterian traditions which are as follows:

- Mousterian of Acheulian tradition
- Typical Mousterian
- Denticulated Mousterian

- Charentian Mousterian

Use of Cave and Control Over Fire: In order to protect from cold environment of fourth glaciation period, Neanderthal man chose to live in caves. Before them Peking man or Chinese man had started living in caves. Neanderthal man started living in caves wherever they found them nearby streams or rivers and where stone fragments were available abundantly for making tools. The tradition of cave dwelling continued in upper Palaeolithic period, too. The period of Neanderthal is also known as early cave dwelling age and the period of upper Palaeolithic is also known as upper cave dwelling age.

Neanderthal man would have entered in cave for dwelling in order to save from dangerous wild animals like mammoth, bear, rhinoceros, etc. These dangerous animals were also trying to capture caves for their protection from severe cold climatic condition of glaciation. Neanderthal could become successful in driving out those dangerous animal from the cave with the help of fire. Neanderthal man knew the use of fire. But, it is very difficult to tell whether they used to light fire or not. Majority of scholars believe that Neanderthals knew the technique of lighting fire by striking stones. The fire was such a useful tool which compelled animals to run away. This was also a useful means of protection from cold. They used to light fire at the entrance place of cave and sleep peacefully. The light of fire had made Neanderthal man stronger than dangerous wild animals. During night time the fire not only provided them with protection, but also light to see in the darkness and energy to cook food. The fire had made their food more tasty. Those materials which were not eaten as raw, they now become favourite food items for them after roasting. The control over fire was also essential for progress of civilisation. Man could be able to melt metals by controlling fire. He was able to make metal tools which made the life very easy. Neanderthal man had contributed a lot to civilisation by establishing control over fire.

Hunting and Gathering: Neanderthal man was completely dependent on nature. They used to consume food materials provided by nature. They did not know to do agriculture. They did not know to domesticate animals for milk, meat and skin. Their food consisted of wild fruits which they used to pluck easily or animals whom they used to hunt individually or collectively.

Wild root, shoot, leaves, flower, fruits, seeds, honey, egg, earthworm, insects and frogs were their common food. They used to catch fishes with their hands from ponds and rivers. They used to collect seaweeds and shells from sea to use as food. They used to hunt small birds probably. They used to hunt small animals individually. But they used to hunt big animals collectively. Animals like mammoth, bear and rhinoceros were associated with Neanderthal age. Neanderthal man had only flake tools. Therefore, they were not able to hunt big animals without encircling them collectively. When big animals were found ill or wounded, they were compelled to move towards water source or ice by their collective efforts. The animal, thus, trapped nearby the water source was killed by them. Only small bones of dead animals have been discovered abundantly from the cave sites of Neanderthal man. But the bones of ribs and backbone have not been discovered. They used to eat the body of the hunted animals at the site where hunting was done. Remaining small parts of the body were brought in the caves for eating. They used to cut them with scrappers. Neanderthals used to get skin hide from the hunting of big animals. They used to take out skin hide with scrappers. They left it in sun for becoming dried. The dried skin hides were used for wearing, covering body and sleeping.

Social Life: Neanderthals used to hunt big animals. Hunting of big animals required collective effort and some kind of organisation. This reveals that they were living in group. Each group had a head. Women and children were more in number than males in the group. Those members who did not obey the head, they were ousted from the group. They had division of labour in the group based on age and sex. Men of the group used to collect food items in day time and gather at a particular place during night. They lived in group to protect themselves from the attack of wild animals. Women and children used to collect stone fragments in order to shape tools. During night the head and other men of the group used to prepare tools from the stone fragments collected by woman and children. The children used to learn the art of tool making from the head and other men of the group. When some boy became quite young, he wanted to become the head of the group. He had to fight with head of the group for this purpose. If he defeated the head, he occupied the post of headship. Now all women, children and men remained under his control. But when he was defeated by the head, he had to leave the group as punishment. Sometimes, he was killed by the group leader.

Death Rituals: In the last phase of his existence, Neanderthal man had started to bury their deads with regards and respects. They used to dig out burial for this purpose. They used to put dead body in that burial prepared for death ritual. Those burials were generally made nearby the place of fire in the cave. They probably had developed the concept that living body possessed heat and dead body possessed cold. They had developed a belief that dead body when buried nearby the fire place would attain rebirth soon. They laid their deads in a special pose and put other materials like tools and food. One Neanderthal youth have been discovered buried in special pose. His head was resting on wrist supported by beads of stone tools in the form of pillo. One hand axe was found nearby his head. A number of small bones were spread there. Perhaps they believed that individuals had their existence in some form even after their death. Dead need food items and tools which he used during his life time. This indicates clearly that barbarous Neanderthals had started thinking on the problem of life and death.

End of Neanderthal: Neanderthal species met their end nearly 40,000 to 35,000 years before today. They met with end by those human species whom we anthropologists call Homo sapiens or true man. As referred to earlier, early Homo sapiens had emerged in Europe during lower Palaeolithic period and Neanderthal man basically was a branch of Homo sapiens. Before this view, a number of scholars believed that Homo sapiens and Neanderthal man were so different physically and mentally that they would not have come in contact with each other. Probably Homo sapiens treated the Neanderthal man differently. Probably they used to hate them because Neanderthals were short-statured and ugly. Therefore, blood contact between early Homo sapiens and Neanderthal man was not possible. Neanderthals became extinct after receiving defeat in war with Homo sapiens. But in recent decades human remains of such species have been discovered from Palestine and Middle Asia that none could deny accepting them in between Neanderthal man and Homo sapiens. Skull remains found from a cave situated nearby Gallillo sea in Palestine and skeletons from caves of Carmel mountain reveal that they were not Neanderthals, rather they were Neanderthaloid. In 1938, remains of a Neanderthaloid child was discovered from Uzbekistan in Russia. The remains of this child shows a mixture of Neanderthals and Homo sapiens. This shows that the blood contact between Neanderthal and Homo sapiens would have taken place.

Remains of Neanderthal Culture: Whether the blood of Neanderthal man is flowing in Homo sapiens or not is a matter of debate. But even today Neanderthal culture is found persisting in Tasmania. When Dutch traders discovered Tasmania in Modern period, they found such a human species there who resembled very much with Neanderthals in their living. Physically this human species belonged to Homo sapiens. This also reveals fact that Neanderthals were also member of Homo sapiens. Physical demerits appeared in them due to environmental change in Europe during middle Palaeolithic period. But the Tasmanian species were able to maintain their physical structure intact. Due to climatic condition they were isolated from rest part of the world. They also remained outside the civilised world. That is why they remained in the same primitive form in which they existed in middle Palaeolithic period.

10

Evolution of Upper Palaeolithic Man and Culture

Human Species

In previous chapter we have seen that Neanderthal man had established his control over Europe during Fourth Glacial (Wurm) Period. But nearly 35,000 fears ago from today he began to disappear suddenly. In place of Neanderthal, such human species began to emerge whose physical features were very much alike to modern man. Their cranial capacity, teeth, chin, shoulder, feet and hands were same in structure as man possesses these physical features today. Anthropologists have termed those human species as Homo sapiens (modern man or true man). After the emergence of these species, the macro physical evolution of man gets stopped but cultural evolution remains continued till these Homo sapienes did not enter into the stage of metal age civilisation.

Place of Origin of Homo Sapiens

Evidences related to the emergence of Homo sapiens appeared simultaneously from Europe, Northern Africa, Eastern Africa and Asia. Therefore, it is very difficult to say where Homo sapiens appeared for the first time. On the basis of discovery of Piltdown man, some English scholars had considered England as the place of origin of true man or modern man. But when Piltdown man became debatable and proved a great hoax, this view cannot be accepted. Some other scholars believe that when Neanderthal man was inhabiting Europe during Middle Palaeolithic period, early Homo sapiens were also

Fig. 10.1: Rhodesian Man

leading their life in the same manner at early places as Neanderthals were leading. These early places were situated in Asia or Africa probably. From those early places they reached to Europe through Northern Africa. Perhaps major part of Mediterranean sea was dry and northern Africa and Europe were directly connected together. Therefore, they did not face any trouble in reaching to Europe. Rather they migrated and settled in Europe directly. Still some other scholars believe that the dry part of Mediterranean sea itself would have been place of origin of modern man. Some anthropologists lay emphasis upon the evolution of modern man through Neanderthals. But we have seen that early Homo sapiens had emerged during lower Palaeolithic period and Neanderthal man was also a branch of Homo sapiens. Due to adaptation with environmental condition he had developed different physical structure. For example remains of Swansecombe man, Steinheim man and Fontechevade man. These men resembles in physical features with modern man. They have been placed unanimously under early Homo sapien group and are termed as early Homo sapiens. In 1951 C. Coon discovered a skull of Homo sapiens from Iran. Its age was 75,000 years to 10,0000 years from today. On the basis of these fossil evidences, it has been established unanimously that the Homo sapiens existed in upper Palaeolithic period. The discovery of 1,00,000 years old remains of Homo sapiens from Iran bears testimony that Homo sapiens had come to settle in Europe nearly 35,000 years ago from Western Asia. Evidences obtained from Galilee sea and Carmel mountain also present proof in this regard. Here one also finds intermingling of blood of Neanderthal man and early Homo sapiens.

Homo Sapiens of Europe

When Homo sapiens conquered Neanderthal man in Europe, they themselves were divided into several branches. Their three main branches were as follows:

- Cro-magnon man
- Grimaldi man
- Chancelade man

Cro-magnon Man

Fossil remains of Cro-magnon man was discovered in 1868 from France. Five skeletons of Cro-magnon man were discovered. After

Fig. 10.2: Cro-magnon Man

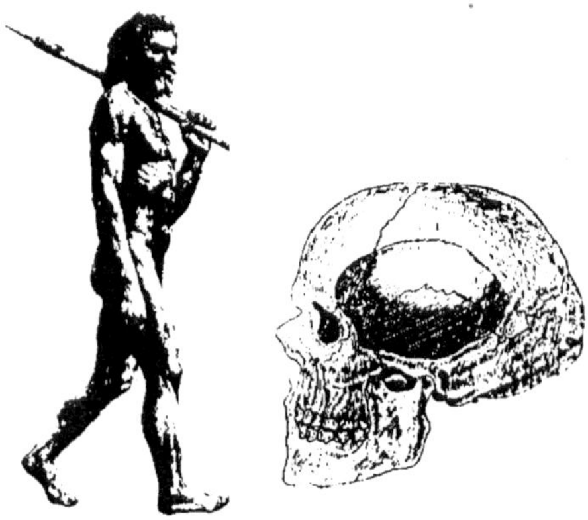

Fig. 10.3: Cro-magnon Man and His Skull

that remains of this man were discovered from several parts of Europe. This man belonged to upper Pleistocene period geologically. They also belonged to upper Palaeolithic culture. Their culture was termed as Aurignacian culture. Their cave drawings, engravings, and artefacts were full of aesthetic sense. They are believed as descendants of Neanderthal man. The chief physical characteristics of Cro-magnon man are as follows:

- Its skull was big and heavy.
- Head length and head breath were 20.3 cm and 15.0 cm respectively.
- The head belonged to dolichocephalic group.
- According to Prof. Keith cranial capacity of this man was 1600 CC but according to Mr. Brown its cranial capacity was 1590 CC.
- Vault was very high.
- Its head was pentagonal in structure.
- Its head was flat and high.
- Supra orbital ridges were wide.
- Occipital bone was elevated.
- Its face was very flat and small.
- Eye orbits were more in width than height. Its structure was rectangular.
- Nasal bone was high.
- Maxillary part shows prognathism.
- Its palate was narrow and slightly elevated.
- Lower jaw was strong but not heavy.
- Sigmoid notch was deep.
- Chin wall well developed.
- Linea aspera was developed in femur.
- Femur and tibia fibula were platymeric and platynemic respectively.
- It possessed long stature and strong muscles.
- Its average height was measured as 168.4 cm.
- The ratio of its hands and feet were like to Negro.

Some people of France like Dordogne may be called as descendants of Cro-magnon. Canary islanders of Guanches resemble very much with Cro-magnon. They also exhibit cultural similarity with Cro-magnon. It is assumed that they have been found from Neolithic to present day.

Grimaldi Man

Skeletal remains of Grimaldi man have been discovered from a cave situated in the village Grimaldi of Italy. On the basis of place of cave, the man was named as Grimaldi man. Among fossil remains, there were skeletons of one 30 years old woman and a 15 years old child. It

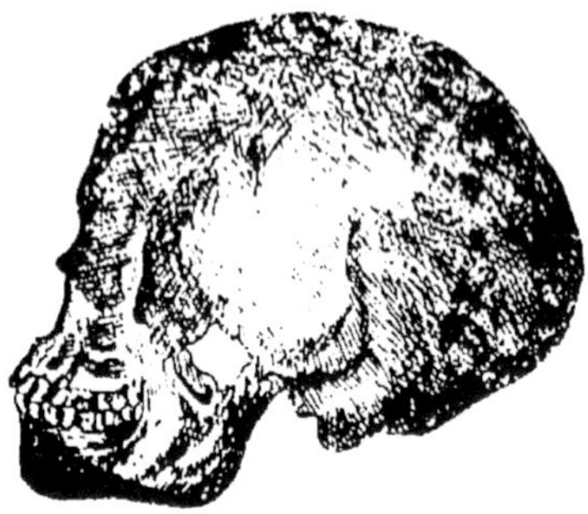

Fig. 10.4: Skull of Grimaldi Man

belonged to upper Pleistocene and upper Palaeolithic period. The chief physical features of Grimaldi man were as follows:

- Its head was long, narrow and high. It belonged to hyper dolichocephalic group.
- Its head height was 11.5 cms.
- Cranial index of woman was 68.5 and child was 69.2.
- The cranial capacities of woman and child were 1265 CC and 1455 CC respectively.
- Elliptical contour was present on head.
- Mastoid process was small.
- Its face was big but not very high.
- Supra orbital ridges were less developed.
- Head was fully developed.
- Eye orbits were big and sub-rectangular.
- Its nose was platyrhine.
- Nasal bridge was small.

- Nasal root was depressed.
- Face shows prognathism.
- Palate was very deep, long and narrow.
- Alveolar border was U shaped and resembled with Negroid and Australoid people.
- Lower part of nasal aperture was sharpen and resembled with Negro.
- Jaws were strong.
- Chin less developed.
- Teeth were big.
- Molar teeth of upper jaw possessed 4 developed cusps and molar of lower jaw showed 5 developed cusps.
- Its teeth resembles with Australian people and head resembles with Negro people.
- Stature of woman was 5'3". But the stature of boy was 5'4".
- Its proportion of hands and feet was similar to Negro people.

Grimaldi man exhibits physical characters which are very much similar to Negro people. They can also be compared with Bushman and Hottentots.

Chancelade man

In 1888, a human skeletal remain was discovered from a cave situated nearby the place Chancelade in France. It was named as Chancelade man. The hands were placed at chest and knees were placed nearby

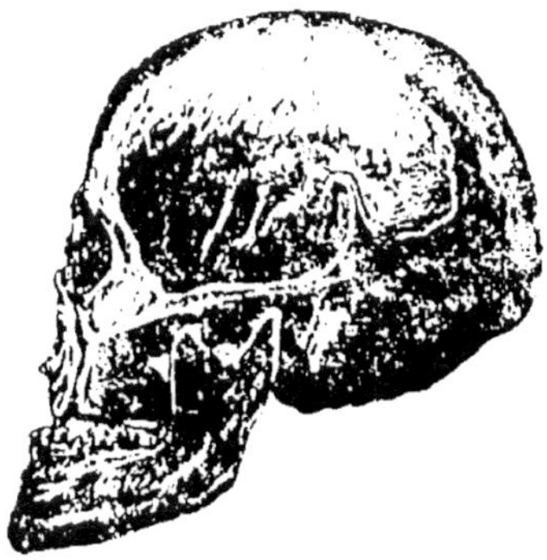

Fig. 10.5: Skull of Chancelade Man

the mouth. It would have been living in Magdalenian period. Geological time scale is upper Pleistocene period and pre-historic timetable is upper Palaeolithic period. It had low stature. No one was more than 5 ft in length. It had heavy body and big skull. It resembled with Eskimos of Greenland Alaska in physical features. The chief characteristics of Chancelade Man were as follows:

- The head length and head breadth of Chancelade man were 19.4 cm and 18.75 cm respectively.
- Its cranial index was 70.9. It belonged to Dolichocephalic type.
- Its cranial capacity was 1530 CC to 1710 CC.
- Brain vault was high.
- Supra orbital ridges were less elevated.
- Head was found in vertical position.
- Parietal tuberso sifies less developed.
- Mastoid process fully developed.
- Face was long and flat.
- Chick bones fully developed.
- Nose leptorrhine type.
- Eye orbits big and quadrilateral.
- Prognathism present.
- Lower jaw strong but narrow in structure.
- Chin developed and showing speciality.
- The age found from the remains was 66 years to 65 years.
- Its muscles were very strong.
- Tibia bone was platynemic
- Femur was platymeric.
- Its palm resembled with Neanderthal man.

Tool: Human species which came to settle in Europe after Neanderthals were more wise and intelligent. Their aesthetic sentiments were more developed than any human species of Palaeolithic age. Their life was more complex than previous human species. That is why they needed different types of tools to fulfil their needs. They could not remain dependent only on stones for making their tools. Rather they used horns, teeth, jaw and other bones abundantly for this purpose. Tools made from these new material bones were polished to make beautiful desired shape. This technique of polishing and grinding was used abundantly for making stone tools in Neolithic period. They also started making stone tools by new techniques. Till middle Palaeolithic period, stone tools were made up largely of core and flakes. But upper Palaeolithic man started making blade tools. This parallel flakes are called blades. Thus, all flakes are not blades. It was easy to make such tools. These tools were also required for their artists to engrave pictures. Burin or gravel occupies significant place among all upper Palaeolithic tool.

Culture: Archaeological anthropologists have divided upper Palaeolithic culture (Fig. 10.6, 10.7 and 10.8) into three periods which are as follows:

1. Aurignacian culture.
2. Solutrean culture.
3. Magdalenian culture.

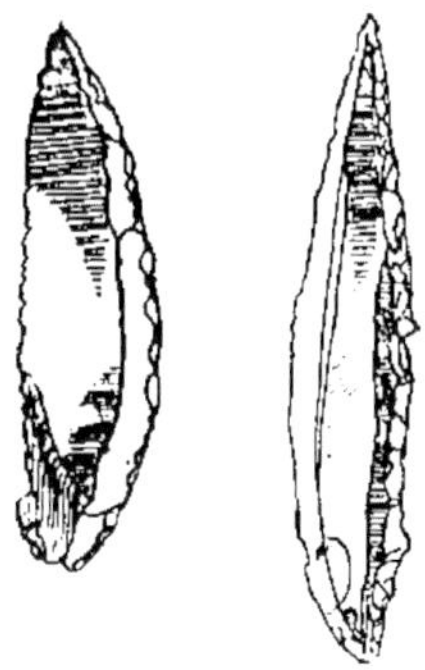

Fig. 10.6: Aurignacian Solutrean Tools

Fig. 10.7: Solutrean Tools

It is very difficult to establish the relationship of man of that period to these cultures. It appears that the same species was related to two-three cultures. Similarly one culture was related to several human species. The chronology of this culture is also unknown. They chronology can be established in very simple form.

Fig. 10.8: Magdalenian Tools

Aurignacian culture: The first culture of upper Palaeolithic period was discovered from a cave named Aurignac of France. On the basis of site name, the culture was named as Aurignacian. This culture is again divided into three sub-periods:

- Upper Aurignacian or Chatelperonian
- Middle Aurignacian
- Late Aurignacian or Gravetian or Perigordian

In Western Europe, the mixed form of Chatelperonian and Gravetian is called Perigordian.

The rise of Aurignacian culture probably took place in Western Asia, but in the end phase of Mousterian culture it spread gradually towards East and Middle Europe, Italy, South France, Northern Spain and England. One also finds tools resembling with Aurignacian culture in Palestine, East Africa, Cyberia, North China and South India. The tools of this culture include polished bone tools such as point, pin, awls, spearhead, core, scrappers and beautiful knife of blade.

Solutrean culture: One finds the existence of blade tools in this culture. The tools have been found spread from Eastern Spain to Black sea. These tools occupy significant place for their beautifulness. These tools were not made by polishing, but they are like thin edged knife. Laurel and Willow leaf shaped spearhead tools occupied prominent place in this culture. Awl and spear of horns of deer were made in this period. They also prepared needles of bones.

Magdalenian culture: This culture is named after a site named Le Magdale in France. This is the finest culture of entire Palaeolithic period. Stone tools of this culture had become gradually smaller and smaller. Majority of tools were prepared from blades. Deer horns, horns of other animals, and bones of animals were also used for making tools in this period. Harpoons made up of bone occupied significant place. Harpoons were used for fishing whales. Spears of horns were also made. Bone awls and needles were also made in this period. Such beautiful needles do not appear even in historical period till 14th and 15th B.C.

The tools of this period were given artistic look by engraving varieties of pictures. Magdalenian man had also invented such an instrument by which spear head could be thrown far away. With this invented tool, fast running animals could be hunted easily by an individual.

All these three upper Palaeolithic cultures were flourishing in Europe and Asia. African cultures contemporary to these European and Asian cultures were Aterian and Caspian.

Aterian culture

Aterian culture flourished in North Africa. In this culture, tools belonged to Mousterian tradition. The builders of this culture knew to make arrow heads whose edges were prepared on both sides. Arrow head was probably first machine prepared by man The credit of inventing such arrow head for the first time goes to African man. With the help of this tool, distantly located animals or bird could be hunted.

Caspian culture

Like Aurignacian, Solutrean and Magdalenian cultures of Euresia, Caspian culture is also a blade culture. This culture had its spread up to South Spain, Italy and North Africa. The builders of this culture were also aware of bow and arrow. Caspian man had also invented Microliths in upper Palaeolithic period. Microliths got more popularity in Mesolithic period.

Social life

Upper Palaeolithic man lived in group. Each group had a head. In group, the number of women and children were more than men. Children were probably known by their mothers. The concept of affinity and consanguinity had not developed. The concept of exogamy was also not developed. The capture of women would have been prevalent. Division of labour based on age and sex was prevalent. Women had to rear and look after children besides collection of food. Children and aged used to assist them. Men remained busy in hunting and fishing. Collection of stones for tool making was done by women and children. Tools were fashioned by men and children learnt to make tool from them. They lived in cave or hut. Cooking was done generally by women. Scrapping of skin was done by males. But drying skin in sun and the work of sewing was done by females. The form of society would have been matriarchal. For the purpose of hunting collectively social organisation had developed. The head of the group managed hunting on collective basis. Animals hunted were distributed equally. But some special share was given to head, priest and owner of the tool.

Economic life

House and cloth: When Homo sapiens conquered Neanderthals in Europe, the climate had become hot than before. Therefore, they could live below sky easily. But the cold climate of fourth glaciation had not come to an end completely. Thus, they used to live also in caves. This period is also known as late cave age. Where caves were not available, they erected tent with the help of dried animal skin for the protection from cold waves. They used to dig out hole in the earth and erected tent over the hole. Perhaps they had also developed the technique of erecting huts. Pictures made by them in caves reveals some structure of huts. Wood was not available to them. From scrapping tools and needles it appears that they used to sew skin and wear them as cloth. The pictures of women drawn by upper Palaeolithic man in East Spain bears testimony in this regard.

Food: Like his ancestors upper Palaeolithic man did not know to cultivate land and do animal husbandry. His livelihood was still dependent upon gathering, hunting and fishing. In these economic activities he was more efficient than Neanderthal man. He used to collect more food materials with the help of bow and arrow, a tool invented by him in this period. He used harpoons for fishing big fishers. He used needle for catching general fishes. He knew technique to trap big animals by digging hole nearby water source where big animals used to come for drinking water. He had also become aware of change in season or climate. He was also aware of nature of wild animals and their movements in different regions with change in climate. Thus, he was aware of place and time of migration of animals. His knowledge was helpful in hunting. He used to pitch his tent in the way of migration of animals and hunt them easily. Their hunt included mammoth, rein deer, yak, rhinoceros, zebra, wild horse, wild buffalo, etc. Bones of those animals have been found abundantly along with the remains of upper Palaeolithic man. He was also aware of varieties of plant species and their nature, and availability of root, leaf, stem, flower, fruits and seeds. Thus, the economy of upper Palaeolithic man was completely dependent on gathering, hunting and fishing.

The Earliest Specialist

The hunting of big animals like mammoth, wild buffaloes, wild horses, rein deer, etc. required cooperation and organisation. For this purpose, upper Palaeolithic man lived always in group. Some kind of division of labour had also emerged. In his group one man was such whose work other men of the group were unable to do. That individual was artist. He used to decorate cave house with pictures drawn or engraved by him. These pictures possessed religious value, too. So, in some sense he can be called as priest. He remained engaged in the work of art and priesthood throughout the day. He did not find time for performing other kind of work. Therefore, his needs were fulfilled by the group members. We can say that the artist-priest of upper Palaeolithic man was the earliest specialist of the world.

Reciprocal contact

The upper Palaeolithic human groups were self-sufficient. Their economy had not become so complex that one group was dependent upon other for fulfilment of needs. The tools required for hunting, gathering and fishing were prepared by the group members. Each group knew well about the techniques by which blade and bone tools were made. For this purpose one group was not dependent upon other group. But this does not mean that reciprocal exchange of goods between groups was not done. There are evidences supporting the exchange of goods between groups. Different types of shells and fish bones of Mediterranean sea have been discovered along with remains of Magdalenian culture in France. This reveals that upper Palaeolithic human groups did not live in isolation. The groups living on sea coast used to exchange shells and fishes with groups of middle France.

Religious belief

Upper Palaeolithic man was fond of art and painting. The caves where upper Palaeolithic artists used to make pictures were also a kind of temple for them. Seeing of those pictures before going for gathering, hunting and fishing was taken as auspicious thing. From this point of view, artists were priests of these cave temples. Priests artists had control over magic. They had control over the soul of animals. With the help of animal souls, food materials for the group were made easily available. It is evident that such artist-priests were powerful and influential in the group. They can be termed as protector of religious beliefs of upper Palaeolithic man. Woman images made by them perhaps were symbol of mother worship. Mother power would have been worshipped in this period in one way or the other. Images were also engraved on bone tools. Images were perhaps engraved to make the tool more powerful. Small images were perhaps fetish. They were used for protection of the group. But it is not essential that all artefacts and images were full of religious sentiments. Some of them would have been an attempt towards expression of beautifulness.

Palaeolithic man possessed belief in life after death (other world)

Their belief in other world was superior to Neanderthal man. They not only buried their deads, but also placed food materials, tools and ornaments along with dead body in the burial. They used to paint the dead body with red colour. Red colour is symbol of blood. They had concept that redness of life returns when the dead body is coloured with red colour.

Knowledge science

Palaeolithic man possessed several knowledge indirectly and laid the foundation of future knowledge and science. He had studied the structure of animals and plants in detail before engraving them in the form of pictures of making images. In the study of physical structure of animals they had progressed much. They had made pictures of different types of fishes. It means that they possessed knowledge of varieties of fish. They also had knowledge about growing of plant species. In one picture, the heart of elephant has been made at right place. In the matter of food materials, they enhanced the knowledge established by Neanderthal man. They had acquired knowledge about poisonous and non-poisonous nature of varieties of food. They had acquired knowledge about the availability of food materials according to season. They had also acquired knowledge about the migration and migration root of the animals. These matters were related to knowledge and science of upper Palaeolithic man.

Art

Reporting about the Palaeolithic art was done for the first time by Sautuola in 1880 through a pamphlet. Since then many discoveries of art specimens have been made. The upper Palaeolithic period is characterised by dramatic appearance of art both on movable objects like engraved bone tools and immovable objects like wall and ceiling The former has been termed as art mobiliser by French historians and home art by their English counterparts. The latter is called as cave art.

Cave art

Art work executed by lines engraved on either of the two kinds of objects is known as engravings. The art work done by paints involving pigments like ochres, minerals, burnt bone, etc. is called painting. The carvings so engraved that the decoration stands out of the panel is known as base-relief. All the three varieties of art are evidenced by the upper-Palaeolithic finds.

Ornaments and body decoration

Varieties of shells and elephant teeth exchanged from distantly located places were used for making ornaments. Upper Palaeolithic man was lover of beauty. He prepared varieties of ornaments to decorate his body. He used to carve different types of images and designs on his ornaments. He used to colour his dead by red dyes. This reveals that he used to decorate his body during lifetime also. Even today some primitive groups decorate their body with different types of dyes.

Painting and drawing

The love of beauty and attraction towards colour of upper Palaeolithic man gave rise to painting and drawing. Being savage, he had progressed much in the field of art. His progress in the field of art was really surprising and alround. He did not do only wall painting and drawing but he carved beautiful images on his bone and horn tools. He also made images from soil. He used to engrave animal pictures on the handle or other parts of bone and horn tools. He also made designs on bone and horn implements. Perhaps he used carvings on skin cloth. Pictures made in low relief of stone fragments have also been found.

In Aurignacian period small images were made by grinding and mixing elephant teeth, stone and soil. Such images have been discovered from Egypt, Creat, Austria, Italy, France and Spain. Women images have been termed by archaeological anthropologists as Venus. (Fig. 10.9). The head of Venus has been shown very short. Venus are probably related to mother power. In place of hairs in head, some lines have been drawn. Stomach, thigh and breast have been shown comparatively big. It seems that the artist had attempted to draw a picture of pregnant woman. Venus are not beautiful from the point of view of beauty. Venus of later period are beautiful and attractive. An image carved on elephant teeth, the artist has succeeded in developing the hair style of a girl (Fig. 10.13).

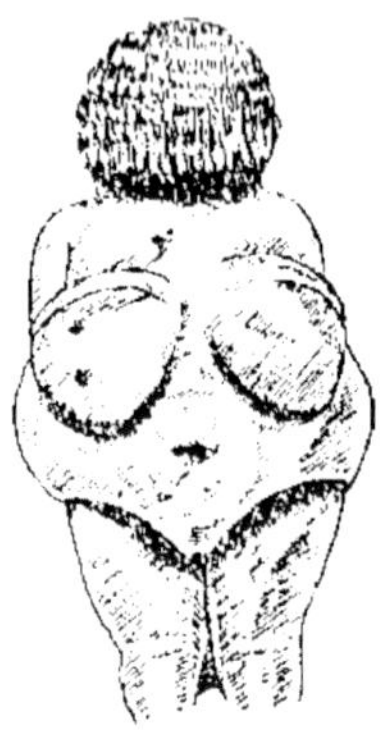

Fig. 10.9: Venus of Aurignacian Period

Early drawing

The early drawings of upper Palaeolithic man would have been like children-drawing of today. In those early drawing only two feet of quardupedal animal (one forward and one backward) were shown. It seems as if the shadow of the animal has been small and lines have been drawn all around. This period was the beginning period of drawing in the world history. That is why they were not successful in solving the basic problem related to drawing. At the time of drawing the picture of a thing, one has to consider length, breath and thickness. The soil mud consists of all these things. It has length, breath and thickness. But at the time of drawing picture on wall or paper, the artist has only length and breath, but not thickness. One can draw geometric picture easily on paper or wall, but to draw the picture of man and animal on paper and wall is difficult because wall or paper does not have thickness; which is required in drawing picture of animals or man. Today this work looks easy, but it was very difficult for upper Palaeolithic man who had to solve this problem by his own attempt and experience. The importance of this invention was not only fruitful in the field of art, but also in the field of science and literature. It is because our knowledge and science is dependent upon the invention of script. Development of script was impossible without the origin of drawing.

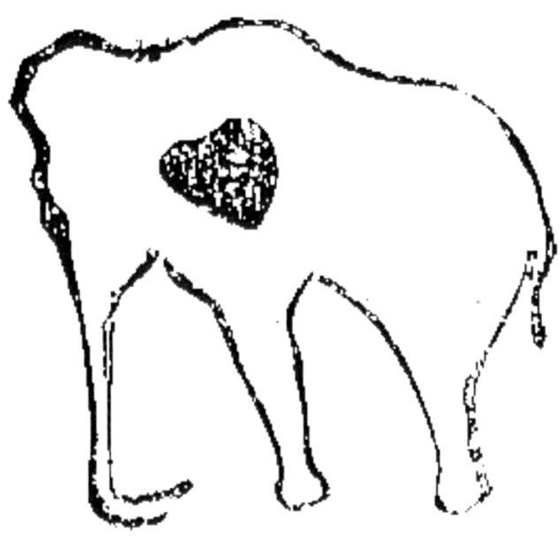

Fig. 10.10: Mammoth of Aurignacian Period

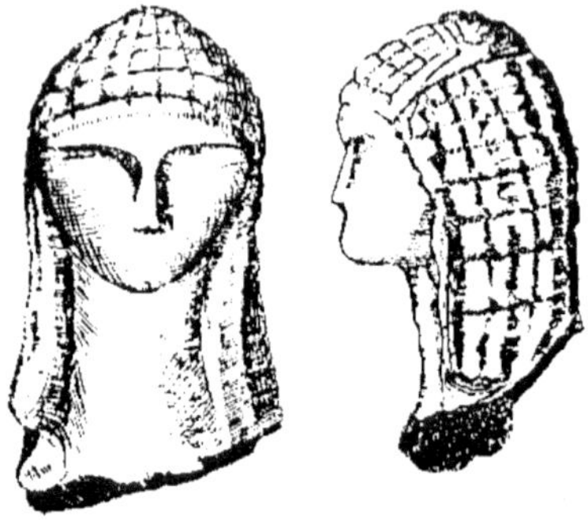

Fig. 10.13: Images Made on Elephant Tooth of Magdalenian Period

Magdalenian Drawing

When upper Palaeolithic man conquered over problem related to drawing, his progress become easy. Gradually his technique of drawing went on improving and beauty of artefacts also went on increasing. By the period of Magdalenian, the technique and beauty of upper Palaeolithic art became so refined that the modern artists also feel pride on that ancestral heritage. The best example of drawing of upper Palaeolithic man has been obtained from the wall and ceiling of pre-historic caves of Altamira a place in North Spain in 1879. A wild buffalo's picture made up with the help of four types of colour occupies most significant place in this regard. This is the best example of art not only of Magdalenian period, but for the entire period of Palaeolithic man. Some pictures are such that they can be termed as suggestive picture. In a picture the band of rein deer has been shown. In this band, one rein deer is in the back side and three rein deer in the front side have been depicted. Others have been represented by lines only the artist has received full success in his attempt here. Some beautiful picture have also been obtained from East Spain. Some of them depict hunting scene. Human pictures were also depicted. This is the speciality of this period. The artists used natural dyes in making pictures. First of all powder of dyes were made and mixed in animal fat for the purpose of drawing. Black, red, yellow and white colour have been used frequently in making pictures. The colours used by them are same till date as they were at the time of making. It is very difficult to say whether they used brush for this purpose or not. It is quite possible that they would have been using brush for this purpose because they had availability of hairs abundantly.

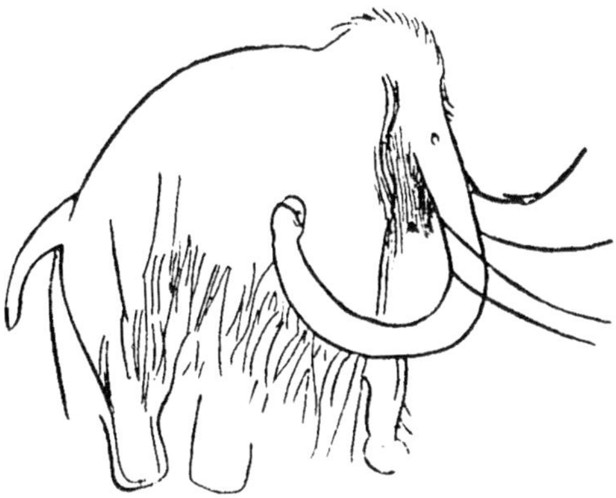

Fig. 10.14: Mammoth of Upper Palaeolithic Period

Styles of cave art

The animals in cave art are almost invariably shown in profile, depicting the three-dimensional animals on two-dimensional canvas. They are more often done in realistic or naturalistic fashion rather being stylistic in their representation. These murals (wall picture) are commonly executed over some naturalistic shape or feature on rock walls. For instance, if there is a natural hole in the cave wall, an animal has been so painted around the hole that the hole represents the eye of the animal. Engravings and relief are also repeated on clay and some slabs found near and around the cave site. The paintings have been done with pigments prepared by grinding naturally occurring orches, mineral oxides and powder mixed with fat, urine, blood etc. served as the paint. To execute the paintings black, red and occasionally yellow pigments were mixed or used independently to draw the outlines of animals. Sometimes, only a series of carefully made dots have been painted to depict outline of the animals. The body of animals are usually left blank, the whole body profile being depicted by a single line. The pictures range in size from quite small ones to very large ones. The pictures, though drawn in two-dimensions always tend to project motion. The position of their legs suggests as if they are never standstill. Sometimes, they are crouching and getting ready to spring the running motion is effectively indicated by showing multiple tails and feet. The paintings in line and also engravings are more often found superimposed over the other. These suggest that many artists had visited the cave site at different times and engraved their own drawings over the previous drawings. Sometimes, they had taken some parts of the earlier drawings to form the profile of the super imposing piece. Panels showing hunting scenes with Javelin marks on the body of animal drawn are also not uncommon. Many caves have human hands imprinted on the walls. These are either made by dipping the hands in paints and then imprinting it against a cave wall or bare hand was pressed against cave wall and then colour was applied around it so that the hand print was left in the form of unpainted part. Many of these hand prints show one or two mutilated finger. The females of both animal and human species were shown with bellies distended by pregnancy.

So far 120 cave sites containing relics of the Palaeolithic art have been discovered and many more are being searched for. Most of these art works are done deep into the caves. Sometimes these are found 200 yards from the mouth of the cave on the ceiling or wall of such places where one can reach only by crawling on the ground with the roof touching the shoulder.

Motives of arts

Scholars on the basis of study of drawing and paintings of upper Palaeolithic man in cave sites have attempted to describe the motives or objectives behind the drawing and painting of such arts in those days. Some scholars have opined that the paintings and drawing of upper Palaeolithic arts were pure expression of artistic feelings. Some other scholars were of view that the main motive behind drawings of upper Palaeolithic man was to decorate tools and caves where they lived. But we have some such facts on the basis of which these opinions cannot be accepted.

These drawings are available in cave at such places where there were deep darkness. Even today these places are in the same state of darkness during day time. The then artists would have burnt fats kept in stone cup or skull to get the cave lighted for the purpose of drawings. If the motive of artists would have been only to beautify the cave where they lived, then they would have chosen the places situated nearby the mouth or entrance of the cave for this purpose instead of visiting deep dark and troublesome places in the cave.

Some drawings have been made at such places where artists would have occupied seat with great trouble. It is clear that in order to decorate the cave or to express the feeling of beauty there was no need to bear such trouble.

Although there was sufficient place available in cave, but new drawings were imposed upon old ones. Superimpositions were done even on contemporary drawings. This also shows clearly that the motive of artists behind drawings was not to decorate the cave or not to express the sentiments of beauty.

Some anthropologists have opined that the motive behind the drawings and paintings of upper Palaeolithic artists was related to religious beliefs and tool. This fact is important that in these pictures animals like mammoth, rein deer, bear wild buffalo and wild horses occupied significant place. These animals were also significant for upper Palaeolithic man. He did not eat only meat of these animals, but he used to get bones for making tools and skin for tent and clothes. Some pictures decipher hunting scenes. Some pictures show spear head inserted in the stomach of animal. They held view that hunting of picture of an animal before proceeding for hunting would bring success certainly. It is because they held view that the soul of animal hunted became captive in the picture. Anthropologists call such practices as sympathetic magic. The artists would have made picture of the animal to be hunted. They would have to show these pictures to their group members. The members of the group would have developed confidence for hunting. This process was not less than magic for primitives.

Achievement of Palaeolithic Man

Palaeolithic period is that long period of human history in which man was attempting to maintain his existence by establishing victory over other animals. Economically he was dependent upon nature. Nature fulfilled his entire need. He lived in forest or cave. He used to eat root, shoot, leaves, flower, fruits, seeds available in the forest. He used to hunt animals to eat meat. He did fishing. His tools were made of stone, bone and teeth. As he was solely dependent upon nature, therefore, his progress was very slow, but undoubtedly, he made progress. The tools of man were made from stone, bone, teeth, horns right from early period, but their type, utility and beauty gradually went on growing.

In this age man established his control over fire. The control of man over fire made his progress more easy. Now he began to eat roasted food materials. Thus, his food became more tasty. He began to protect himself from cold waves, darkness and wild animals with the help of fire. The control over fire also opened the way of metal work in future.

As man was dependent upon nature for very long period, he gradually began to experience and study nature and natural events. He gradually experienced which plants and animals were available in which environment. He also experienced about the utility of plants, animals, rivers, ponds, caves, seas etc. for his living and existence. His such experience led the foundation of knowledge and science in latter periods.

The Palaeolithic man got his great success in the field of art. Even today we do not find such persons who are able to draw pictures without any kind of training as Magdalenian used to draw.

In spite of all these achievements, Palaeolithic man remained quite unsuccessful in the field of economy. After attaining a limit, his progress was retarded.

11

Evolution of Mesolithic Man and Culture

Human Species

We have seen in previous chapter that man had evolved in the form of Homo sapiens or true man or modern man in upper Palaeolithic period. After evolution of man as Homo sapiens, no further macro-evolution occurred in the physical type of man. But his cultural evolution continued till he did not enter into the age of metal. However, micro-evolution is still going on and will continue to go in human population.

Transition phase

Undoubtedly man had achieved progress in the typology of tools and art in upper Palaeolithic period, but he was unable to get success in the field of economy. Although he was successful in solving the food problem to some extent by hunting big animals like mammoths, reindeers, wild buffaloes and wild horses, but he was fully dependent upon nature for his survival. He had got sometime to show his ability in the field of art. He could not achieve knowledge of agriculture and animal husbandry. In other words he was not successful in compelling the earth for yielding more and more food materials for his existence. Agriculture and animal husbandry, both these inventions were made in Neolithic period. In many regions of the world Neolithic age started just after Palaeolithic age. But in Europe and some other parts of the world, human culture witnessed a transition phase. This transition phase is termed as Mesolithic age or middle stone age.

Geographical change

Geologically, Mesolithic age is a transition phase of Pleistocene and Holocene. After Magdalenian age, a phenomenal change occurred in the geography of Europe and Asia. Mediterranean sea which was in the form of two lakes now got filled up and achieved his present form. The big sea located in middle Asia began to dry and changed into Caspian sea, Black sea and lakes of middle Asia. Spain from Africa, England from Europe and Arab island from Egypt were separated. India also assumed her present form in the same period. A drastic change in the environment of these continents took place. The *savana* (green fields) of Western Asia and North Western India assumed the form of deserts. The glacial cold ended completely in Europe. Now Northern Europe was covered with forests. Palaeolithic animals found living in cold climate began to migrate gradually towards north. Thus, mammoths and reindeers moved towards north. Animals adapted with warm climate began to appear. These animals were quite adapted to cope with the new environment. Along with new animal species, new branches of Homo sapiens also appeared. New branches of Homo sapiens defeated Cro-magnon man in Europe. These geographical changes brought direct effect upon the life of Mesolithic man. He had to adapt himself with the changed environment. From the point of view of short development period, Mesolithic period appears lower than Magdalenian, but from the point of view of long period of development the Mesolithic period shows progress. This period provided the man with occasion to prepare his ground for Neolithic revolution.

Life of Mesolithic Man

Like upper Palaeolithic man the main food material of Mesolithic man was meat obtained from hunting of animals. But one finds change in the animals hunted and technique of hunting completely. The upper Palaeolithic man used to hunt mammoths, reindeers, wild buffaloes and wild horses. He used to do hunting of these animals collectively. Therefore, man lived in this age in big groups. But in Mesolithic period, the number of these animals got reduced. Now man did not need to live in big group. He began to live in small group. He used to hunt deer, antler, hare or rabbit. He had now invented bow, arrow and microliths.

With these tools he was capable of hunting alone or in small group. That is why man seems spread in small groups in Europe.

In this period, man learnt to take cooperation of dog in hunting. The dog is the oldest friend animal of man. It was the first animal domesticated by man. He used to seek help of dogs in hunting deers and hares. For the act of cooperation, dog used to get a share in the meat of animal hunted. In course of time man learnt to use dog for different purposes. On the basis of domestication of dog, he learnt to domesticate other animals. This gradually led to domestication of animals in Neolithic period.

Art

Mesolithic man like upper Palaeolithic man used to live in cave and huts but he did not possess interest in decorating them. It is true that he had love with dyes. But he did not use them for decorating walls, floor or ceiling of caves. He had made simple pictures on small fragments. Probably he had used symbol-pictures. By this time the symbols of some things were definitely evolved. The artists used to express his art feelings not by drawing pictures but only by some lines. The artist had given certain meaning to these lines. The members of the group knew the meaning of those symbols. Thus, they used to save time and labour by this technique. This technique reveals the decrease in love for beauty but increase in intellectual progress. This led to the invention of script in future.

Fig. 11.1: Designs Made on Stones of Mesolithic Period

Microliths and Culture: One finds the tendency of making tools small in upper Palaeolithic. Gravetian period in France, Solutrean period in East Spain and Caspian period in North Africa show such tools whose size is very small and shape is geometric. These tools were called microlithic tools or microliths. All cultures of Mesolithic period exhibit the production of microliths and microlithic tools. Microliths were fitted in wood or bone handle to make toothed tools. At many places, this tradition continued in Neolithic age and Bronze age.

Fig. 11.2: Microlithic Tools

Azilian Culture

This is the oldest Mesolithic culture of Europe. This culture has been named after a place Le Masd'Azil in France. This culture developed in those regions where Magdalenian culture was flourishing. The creator of this culture used to live in caves. They are famous for their pictorial stone fragments and small harpoons which had a hole. The stone tools of Azilians are like side scrappers and burin, but their shape and size are very small comparatively.

Tardenoisian Culture

This culture seems to related to Azilian culture in the beginning. But it became different later on. The creator of this culture used to make geometric microliths. They used to fit microliths in bone handles. They acted as harpoons. They had also made microburins. But very few bone tools have been discovered.

Asturian Culture

This culture flourished in North-Western regions of Spain and Portugal. The creators of this culture were dependent upon shell and fish for their existence. Their stone implements resembled with Eoliths.

Kitchen-Midden Culture

This culture flourished nearly 10,000 years ago. The remains of this culture have been discovered from France, Portugal, Brazil, Japan, Manchuria and Denmark. This culture is known as kitchen-midden in Denmark. The remains discovered of this culture consisted of sea animals like tortoise, fishes, snails, bones of animals, horns and stone implements.

Maglemosian

In late Mesolithic period cold vanished in the regions of South-Sweden and Norway. As a result of this, the descendants of Palaeolithic man came to settle down here. The early tools of this culture are like Aurignacian and Magdalenian culture. But later on, a different culture appeared which was termedas Maglemosian. The creator of this culture used to make harpoons and needle from bones. They made axe from the horns of reindeer. They knew to make geometric pictures on tools.

Time Span of Mesolithic Culture

It is very difficult to decide the time span of Mesolithic period than Palaeolithic period. The Palaeolithic period had continued for a very long time, but the progress made by man was very slow. At that time there was no cultural difference in different regions inhabited by Palaeolithic man. But the process of progress became faster in Mesolithic period. In this period cultural differences appeared in different regions of the world occupied by man. In some regions of the world Palaeolithic period ended soon, but in some regions it ended later on. In Mesopotamia, Mesolithic culture started appearing around 18,000 years B.C., but in Denmark Palaeolithic period continued till 8,000 years B.C. The Mesolithic period also witnessed end in different times in different regions of the world. In Western Asia man had started cultivation and domestication of animals 7,000 years to 6,000 years B.C., but man of Europe became acquainted with agriculture and domestication of animals after several thousand years.

12

Evolution of Neolithic Man and Culture

When in Europe lands were being covered by forests in Mesolithic period and descendants of Homo sapiens were trying to adapt themselves with the changed environment, Western Asia and Northern Africa were witnessing important geographical changes. These changes influenced the lifestyle of mankind. Till now man was dependent fully on nature to feed his stomach. In this age he compelled the nature particularly land to yield food materials. Thus, he began to produce food materials by agriculture and domestication of animals. He started making boat from the wood of forest. He also made implements for doing agriculture. Thus, man made inventions of art of pottery, carpentry and weaving. For these purposes, he required more sharp and efficient tools and implements. He began to make polished stone implements for this purpose. This period of cultural activity of man is termed as Neolithic or new stone age by archaeological anthropologists.

Date of Neolithic Period

There is no doubt that Neolithic period began in the Holocene period. Till date we have no evidence by which we can say that Neolithic culture evolved in Pleistocene period. Evidences available from Eastern Mediterranean bear testimony that Neolithic culture evolved for the first time in this region. In this region man lived at a place for not only century but for thousand years. Their huts made up of soil, stone fragments and wild grass were destroyed frequently. But they used to make another huts at that place. The remains of old huts were buried by new huts. This process had continued for a long period. Gradually, the place of huts occupied shape of a tell. The fields of Greece, Syria, Asia minor, Turkistan, and Iran are filled with such tells. The excavation of these tells reveals several facts about pre-historic and historic periods. The date of remains discovered from these tells has been established as 3,000 years B.C. or one or two century more. Older remains of this belonged to Copper and Bronze age. The oldest remains belonged to Neolithic period.

Fig. 12.1: Neolithic Huts Made in Lakes of Switzerland

Neolithic Villages of Western Asia

The oldest Neolithic colony discovered by archaeological anthropologists is Jeriko village. It lies in the state of Jordan. Its dating by 14 method reveals that man subsisting on gathering, hunting, agriculture and domestication of animals were living here nearly 9000 years ago. So we can say that Neolithic culture began in western Asia nearly a thousand years ago. But inhabitants of this village were not knowing about the production of ceramics and polished stone implements. More or less same situation prevailed here till 6000 years B.C. In the some period same human groups were inhabiting nearby the cave of Carmel mountain in Palestine. They were known as Natufisian. Their stone implements resembled with Mesolithic tools of Europe. But they were able to make a new tool called Sickle. This tool would have been used for cutting grasses. In the Jarmogram of Kurdistan the same condition was prevailing nearly 4750 years ago. But the inhabitants of this village had learnt to make images of soil baked in fire. Their utensils were of wood or stone. The evidence about agriculture and domestication of animals appears for the first time from the Sayal village in Iran. Besides agriculture and domestication of animals, evidences about the inventions of pottery, thread making and weaving were also discovered from this village site. These evidences have been obtained from the first layer of this site. From middle Asia evidences about agriculture, domestication of animals pottery and weaving have been discovered from the oldest layer of a place Anau nearby Astarabad town.

Neolithic Village of Egypt

From a place Fayum on Western bank of the river Nile, remains of 4300 years old have been discovered. The discovered remains included bones of domesticated animals, harpoons for fishing, sickle prepared from microliths fitted in wooden handle, granary in the form of pits, polished axes, pottery, awl and arrow head of flints. Findings of awl and shoulder point reveals that man of this village knew to make clothes by thread making and weaving. Here, granaries are the first example of the world in the direction of attempt to collect food materials for future use. Such granaries were discovered from the place Morimde situated in the North-Western part of delta of Nile river. In excavation; granary has been discovered from every house of each village. Neolithic remains have also been found from the place Alomari located eastwards from Nile river. Such remains have also been found from Tasa in middle Egypt. The inhabitants of these villages were quite acquainted with the agriculture, domestication of animal, pottery and weaving. The remains obtained from the place Badari situated nearby Tasa belonged to some later date. The inhabitants of Badari had trade relation with Syria and they used shells obtained from red sea.

Neolithic Period in Europe

The rise of some traits of Neolithic culture had taken place nearly 10,000 years ago in East-West Asia and Egypt. But the developed form of Neolithic culture came into existence nearly 6,000 years to 7,000 years ago. In Europe, the Neolithic period began nearly 5,000 years to 4,000 years ago. In this content first of all in Creat and Greece, then in middle Europe and Western regions witnessed the industry of agriculture and domestication of animal. In Denmark, Northern

Germany and Sweden Neolithic period began nearly 2,000 years ago. The Neolithic man of middle Europe was known as Denyubian. The excavation of Koln Lindelthal village presented the cultural features of middle Europe Neolithic man.

Neolithic culture had its spread from China to Ireland. Even today this culture has not ended completely. The primitives of Africa, America, New Zealand and other countries were leading Neolithic life till recently. Some primitives are still leading their Neolithic life.

New Inventions

Chief characteristics of Neolithic culture were found in all human species inhabiting the different parts of the world at that time. But there was no similarity in Neolithic culture of different regions. It was because different region did not possess same climate and regional features. For example, at some places jute were used for making cloth, while at some other places cotton threads were used for making cloth. At some places domestication of animal got importance, while at some other places agriculture occupied significance. But in spite of all these regional variations, some chief characteristics of Neolithic culture had emerged. These were as follows:

Agriculture

As we have seen earlier that the originator of Neolithic revolution were not progressive Magdalenian man of upper Palaeolithic period, but comparatively backward human groups of Western Asia, North Eastern Africa and perhaps North-Western India. These places were full of grasses at that time when Palaeolithic period ended completely. In the beginning of Holocene geological period, universal changes took place in environment and climate. Northern Europe was covered by forests suddenly. The climate of these regions became dry more and green lands of forests started changing into deserts. The inhabitants of these places find it difficult to survive on hunting and gathering. This climatic condition forced the man to invent such means which could increase food materials for survival. Males could not achieve much success in this regard. But females got success. Women used to collect root, leaves, flower, fruits and seeds of wild plants and trees. In course of gathering and collecting food materials they observed that the seeds of those plants when buried in mud soil produced more and more seedlings, plants and trees. This phenomenon used to happen in a particular climate and seasons. The observation of pre-historic women on birth, growth and maturity of wild grasses ultimately provided feedback and knowledge about agriculture to Neolithic man. There is no unanimity among the scholars that in which countries of the world agriculture developed for the first time. British diffusionist anthropologists G.E. Smith and W.J. Perry had given this credit to the valley of river Nile in Egypt. In the opinion of Russian scholar Vevilove this credit should go to Afghanistan and North-Western China. Today majority of scholars believe that Natufisian of Palestine were the inventor of agriculture.

Main Crops

The nature had made a number of plants whose seeds are edible for man. For example, wheat, oat, barley, gram, paddy, maize, millet, kandas and potato. Among these, wheat and oat are most energy giver. They are easy to collect. They can be produced in very big amount from very less amount of seed. Its production requires less labour. Cultivators have to render labour at the time of tilling land, sowing and harvesting. Thereafter, the cultivators are free for other kind of work. That is why, these two types of foodgrains have occupied significant place. Since Neolithic period when Neolithic women observed on these plants and seeds, they were in their wild form. But gradually man provided them modern form by cross-breeding.

Agricultural Implements

Neolithic man possessed very few artificial implements for doing agriculture. He was not even aware of plough. He used to till land with hoe (spade). He used to drop seeds even without tilling when environment was favourable. He used to prepare sickles for cutting crops. Sickles were made of microliths fitted in bone or wooden handle. He had to depend nearly for a year, for harvesting of one crop to another crop. That is why he invented granary. Grains were preserved to consume daily for a year. Neolithic granaries have been discovered from Fayum, Merimde and Koln Lindelthal villages. Crushing stones and hearths were invented for crushing these food grains and baking them.

Fig. 12.2: Neolithic Spades

The Problem of Neolithic Agriculture

Due to regional variability of environment, Neolithic peasants made experiment after experiment to make progress related to agriculture. The peasants of Iran and Mesopotamia could not depend only on rain. That is why they evolved artificial means of irrigation. Contrary to this people could remain dependent on rain in Europe. But the lands of these regions used to become non-fertile after two or three consecutive years. Denyabians were aware of this trouble. To overcome on this problem, they left the land barren after tilling two or three years. In some years, when entire lands of the locality became infertile, they shifted to another place for living. This method of agriculture is known as shifting cultivation. Shifting cultivation is

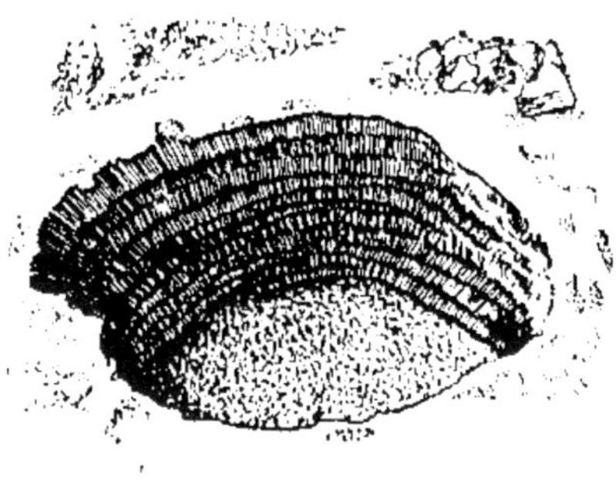

Fig. 12.3: Neolithic Granary from Fayum

still prevalent among a number of tribes of North-East India and Africa. But this method invites various kinds of trouble. That is why Neolithic man began to think upon a plan to omit non-fertility of the soil. Denyabians were first to discover that when wild grasses are allowed to grow in the field and are burnt after cutting, the fertility of land gets returned. Human species of Greece and Balcon invented the method of increasing fertility by cow dung, urine, night soil, etc.

Domestication of Animal

The human species inhabiting western Asia and Mediterranean countries used to do domestication of animals along with agriculture. This industry also came into existence due to climatic change. When rainfall decreased in these regions, grasses became dried and desert started appearing, animals and human species were compelled to migrate towards the places situated nearby Nakhlistan (Mountain valley). A number of animals surviving on grass such as cow, ox, buffalo, sheep, goat, pig, etc. began to wander nearby the places of human habitation. By this time the man had become quite acquainted with these animals. He was also able to know that he could kill these animals easily for getting meat as food material. That is why he did not make any attempt to drive out these animals, rather he allowed them to come closer and closer. He began to provide fodder to them available from his agricultural fields. He began to protect them from dangerous wild animals. Gradually these animals became fully dependent upon him. In this way, domestication of animals came into existence.

Which Came First — Domestication of Animals or Agriculture

On this issue there is no unanimity among the scholars. Majority of scholars believe that at some places domestication of animals was started, while at some other places agriculture was started. But both these industries of Neolithic period originated in the same time. Contrary to this view, there are scholars, whose numbser is very less, who hold opinions that domestication of animals started before the origin of agriculture. But this opinion has not been accepted by British new-evolutionist anthropologist V.G. Childe. He holds view that without the advent of agriculture it was very difficult to solve the problem of fodder. So, agriculture originated first and then domestication of animals started.

Advantage of Domestication of Animals

In Neolithic economy the importance of domestication of animal was not less. With the domestication of animal, the Neolithic man was able to solve his food problem. Now it was not essential to wander in the forest in search of animals for hunting. Now he was able to obtain meat any time by killing his domesticated animals. He used to get skin from the animal. Skin was used for making cloth, tent and pots. So, skin was very useful for Neolithic man. Ornaments, tools and implements were made from the horns of animals. Neolithic man also discovered that grazing of animals in the field was beneficial from the fertility point of view. This practice was useful in getting good harvest from the agricultural fields. Gradually he acquired knowledge about manuring the field by cow dung. He also obtained wool from sheep. This practice gave rise to thread-making and weaving. When he got acquainted with the nature of animals, he knew about milking them. He also knew that milk could be used as food material. He also knew that animals were also useful for carrying loads from one place to another. These were the benefits associated with the industry of domestication of animals in Neolithic period.

Impact of Domestication of Animals

In the beginning domestication of animal did not bring much change in the economic life of pre-historic man. But when number of animals domesticated increased, it brought several problems associated with the domestication of animals. Now he had to graze animals. He had to clear forest for grazing ground. He had to cultivate special crop for the purpose of fodder. There were many such works related to domestication of animals. As a result some man remained always busy only in domestication of animals. They did not find time spare for other kind of works. Domestication of animals became the base of livelihood for them.

There is no doubt the production of food materials from the field started in neo-lithic period. But this does not mean that gathering, hunting and fishing were stopped completely. Gathering, hunting and fishing are still continuing. This practice has also taken the form of business by some human groups. Man associated with hunting is called hunter (Bahelia). The man engaged in fishing is called fisherman. The man engaged in gathering are called forest dwellers, vanyajati, vanjara, etc.

Pottery

Neolithic man was not satisfied by inventing the industries of agriculture and domestication of animals. He was also not satisfied by producing food materials in sufficient quantity. He acquired knowledge to produce such objects which could not be obtained directly from the nature. Among these things were pottery, jute, cotton and wool. He also learnt to make boat. Alongwith the invention of agriculture and domestication of animals. Food materials were produced sufficiently, but there was lack of pots to use them. Till now pots made by man were made up of wood and stone, but these pots were not useful for cooking food materials. To overcome this difficulty, man invented the art of pottery. When and by whom pottery was invented, it is very difficult to say. It is possible that some Neolithic women would have seen that basket covered with mud when burnt in fire, the basket of baked soil remained. This experience led to the invention of pottery. Some scholars believed that pottery had come into existence during Mesolithic period. But it is true that pottery was produced in large scale during Neolithic period.

Fig. 12.4: Neolithic Potteries

Complexity of Potter's Art

Making pottery is a chemical process. Wet soil pots get dissolved in water. They get broken easily after drying in sun, but when wet pottery are heated at 600°C or more in fire, its viscosity gets vanished. They became hard like stone. They can neither be broken easily nor can be dissolved in water. This is the basis of art of pottery. Potter can provide any shape to viscous soil. He can provide a stable shape to pottery by baking in fire. The art of potter was complex from early times. He had to select good quality soil so that it could not receive a gap after drying. It was essential for the potters to recognise the quality of soil required for making pottery. It was also essential to know that pottery needed to dry in sun before baking in fire. It was also not less troublesome job to provide desired shape to potteries.

In the beginning man made pottery of same type as he had made pots from stone and wood. Gradually he invented that different types of pottery could be made from viscous soil. But at that time wheel was not invented. So, he was not able to provide a concrete shape to his imagination. In absence of wheel, he devised ring method to make earthen pots.

When earthen pots are baked in fire, their colour gets changed. The colour depends upon the type of soil, temperature of fire and method of baking. Gradually Neolithic man was abe to know to provide desired shape and colour to the pots. The flow of fire made pots black. To overcome this trouble, the Neolithic man of Asia invented oven. In this oven pots did not get black colour even after receiving smoke and heat at 900°C to 1000°C. In Europe, the benefit of this invention could not be taken earlier than iron age.

Impact of Art of Pottery

It was like a magic for Neolithic man to see viscous soil becoming hard like stone after drying and baking. In case of stone, man was able to make tools of the same size as were stone fragments or their flakes. Similar was the case with bones, horns and teeth. But in case of pottery, there was no such bondage. Man could take help of his imagination in making potteries. Thus, the art of pottery influenced the thinking power of Neolithic man to a great extent.

Art of Thread Making and Weaving

From the Neolithic remains of Egypt and Western Asia, it appears that the art of thread making and weaving was evolved in this age. Thread made, jute made and wool made cloth had started taking the place of clothes made up of skin and leaves. The art of weaving was also not simple. It was very complex. Its invention was imposible without the invention of some other materials and tools. For this purpose such type of material was essential from which chord or thread could be prepared. In Egypt and Europe jute were produced and used for making thread. Second material was cotton. In India cotton had come into use by 3000 B.C. The people of Mesopotamia were using wool for this purpose at the same time. This evidences revealed the cultivation of cotton and jute and domestication of sheep for the purpose of thread making and weaving. Special kind of tools were required for making thread and weaving. Neolithic man made *Charkha* (thread making machine) and *Kargha* (weaving machine). With the help of these two tools and above described materials. Neolithic man was able to make clothes for covering his body.

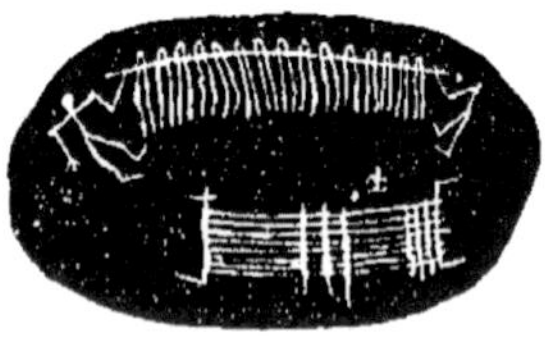

Fig. 12.5: Neolithic Charkha and Kargha for Weaving Thread

Carpentry and Other Tools

During Neolithic period Europe was covered with wood forest. Although the climate of North Africa, Western Asia and North West India was dry but more humid. That is why one does not find scarcity of wood forest in these regions in those early time. In order to use wood and to get success in his new industries of thread making, weaving and pottery. Neolithic man made his stone tools more and more sharp. He began to make sharp edge of his tools by grinding technique. He started polishing his tools by rubbing on another stone. His Neolithic technique of tool making is known as polishing and grinding technique. Among his Neolithic tools, polished axe occupied prominent place. These types of tools were complete absent in Palaeolithic period. With his sharp edged tool axes, he was able to cut and clear the forest. Axes were fitted in wooden handle or horn made handle. He was also able to sawing the wood with the help of saw. Cutting and sawing of wood gave rise to carpentry. Now man began to use wood in making boat, house and other things. Axes assumed the shape of club in its changed form which were used in war. Clubs were of ball shaped in Western Asia, and plate shaped in North Africa and Europe. Besides clubs, other war implements were spear and stone made arrowhead.

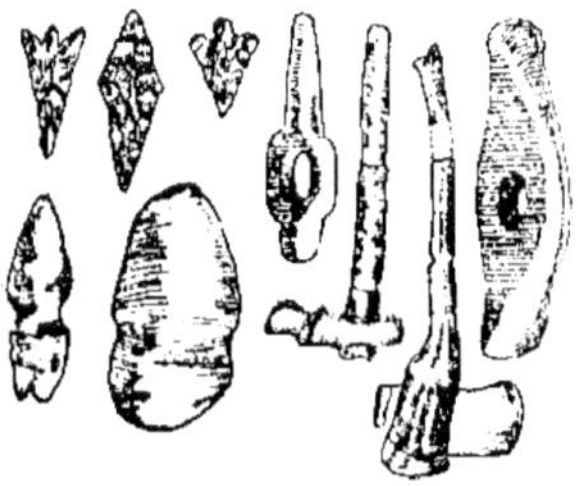

Fig. 12.6: Polished Tools of Neolithic Period

Other Tools

The intellectual level of Neolithic man was superior to Palaeolithic man. Like his ancestors he not only made point, harpoons, needle, pin, awl, spade, shoulder and knife from stone, bone, horn and elephant teeth, but he used his intellectuality in making other tools, too. He made rope to climb upon and boat to cross rivers and lakes. For the purpose of harvesting, he made sickle and for thread making and weaving he made *Takali*, *Charakha* and *Kargha*. Perhaps he used to make drum from soil and wood. Drums were covered with skin of animals. He had learnt to make whistle from *Sarkanda*.

Impact of New Inventions

Earlier we have seen that Neolithic period provided man with occasions of making new and new inventions. About those Neolithic inventions descriptions have been presented just above. Now we are going to deal with the impact of those new inventions on the life of

Neolithic man. We can discuss the impact of new inventions on Neolithic man under the following headings:

- Increase in population
- Beginning of settled life
- Erection of house
- Planning of villages
- Division of labour
- Self-sufficiency of families and villages
- Social organisation
- Magic and religion
- Art
- Death rituals and megaliths
- Sorcery
- Knowledge and science

Increase in population: New inventions made by Neolithic man brought revolutions in human life. For lakhs and lakhs of years Palaeolithic man was completely dependent upon nature for his livelihood and survival. He used to hunt only those animals which were available in the forest. He was able to collect only those root, stem, leaves, flower, fruits and seeds, which were available in the forests. Palaeolithic man had to face the problem of food scarcity. He was not able to get sufficient food materials to provide food to all Palaeolithic population. In case of starvation, infanticide was also practised. This was a strong barrier in population increase. Another point was that when change in environment took place, many species of plants and animals survings in that environment became extinct. Extinction of plant and animal species used to create a great problem of food material for Palaeolithic man. In this situation, it was difficult for man to maintain his survival and population. This event had taken place in the life of Magdalenian man who were well cultured at that time. Magdalenian man had become extinct because he had faced the problem of extinction of plant and animal species in changed environment.

During Neolithic period man acquired for the first time, knowledge about agriculture and domestication of animals. The inventioin of agriculture and domestication of animal had compelled the nature particularly land to yield grains, seeds and fruits as food materials.

Thus, besides food materials available in forest, rivers, lakes and seas, he started getting food materials from earth and domestication of animals. Invention of agriculture and dometication of animals was favourable for increase in population. In this period man used to cultivate more field to feed the increased population. He had also to make increase in animals to be domesticated. The strongest proof of those success in Neolithic period was increase in population. The human groups of Neolithic period were bigger and more in number than Mesolithic and Palaeolithic human groups. In this period man started living in those regions in which he did not live in Mesolithic and Palaeolithic period or his number in those regions was negligible. The number of Palaeolithic human remains are only some hundreds but the number of human remains of Neolithic period are thousands. During Palaeolithic and Mesolithic periods, children were considered as liabilities. They did not help in hunting and gathering, rather used to demand food materials for eating. Thus, they were liability economically. But in Neolithic period, children became assets. They could bring cattle to grazing grounds, look after agriculture and share hands with economic activities of the family. This was also responsible for increase in Neolithic population. Now there was sufficient food materials to feed children. The practice of infanticide was totally stopped in this period.

Beginning of settled life: It is generally believed that man used to wander from place to place in search of food during Palaeolithic period. He was completely barbarous in Palaeolithic and Mesolithic periods, but when he entered into Neolithic period, he began to lead a settled life with invention of agriculture. But this belief is not true. There is no relation between barbarous or hunting man and agriculture or settled man. Magdalenian man was hunter, but he lived in the same cave for generations. During Neolithic period man also had to lead a wandering life like barbarous. This was particularly true for the regions where the fertility of the soil stopped after two or three years of cultivation. In these regions Neolithic man had to lead wandering or barbarous life in spite of entering into the stage of agriculture. But it is true that settled life began in those regions where the fertility of soil was maintained each year due to flooding and man had started manuring the fields by cow dung, urine and other agricultural waste materials.

Erection of house: Palaeolithic man did not know to erect house of any kind. He lived in caves. His dwelling was cave dwelling and he was cave dwellers. But Neolithic man had invented step, pulley and hinge. With these tools he was able to erect a permanent house. In Egypt houses were made of reed. In Western Asia and Europe, houses were erected with mud, wood and unbaked bricks. In Switzerland houses were erected on lakes. The remains of lake dwellings came into existence in 1854 when lakes were dried. Those houses were erected by inserting wooden poles in water of lakes. They had steps for coming and going. Walls of the house were made of wooden framework plastered with mud. Roofs were made of reed, hay and bark. The builders of those houses were definitely skilled carpenter. Such types of houses erected in water have been discovered from France, Ireland, Italy, Russia South and North America and India. These types of water houses are still in existence in Java, Sumatra and New Guinea. From the point of view of protection and cleanliness, these houses were definitely better.

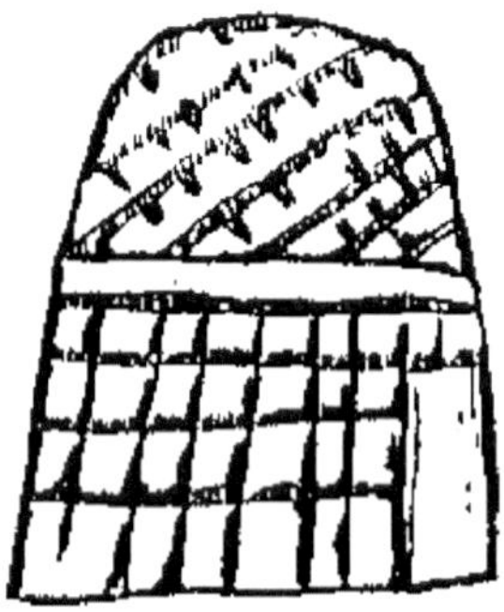

Fig. 12.7: Huts Made of Reed of Neolithic Egypt

Planning of villages: Neolithic man lived in small villages. The area of one village ranged from one acre to ten acre. Jeriko village of Jordan possessed an area of 8 acres. The number of family in villages varied from 8-10 to 30-35. The villagers used to lead a collective life. They had to make road and street collectively. From the point of view of protection, a boundary was erected or a trench was dug out encircling the village. The trench of Jeriko village was 27 feet in width and 5 feet in depth. Boundary or trench was made collectively by the villagers. Houses were erected on both sides of road and street in planned way.

Collective living indicates that Neolithic man had entered into developed stage of social life.

Division of labour: In Neolithic societies, the division of labour based on sex had come into existence. As referred to earlier, majority of inventions were made by pre-historic women. The credit for the invention of agriculture, domestication of animals, pottery, thread making, weaving, etc. goes to Neolithic woman. They had to share maximum family responsibilities. They were given the responsibility to do agriculture, to look after kitchen garden, to crush flour, to cook food, to make thread and to weave clothes. They had also to share the work of ornament making and pottery making. Males used to help women in agriculture and domestication of animals. In off agricultural time, they used to do hunting and fishing. Tools and implements were made by Neolithic men. This reveals clearly that in Neolithic societies women had to perform more work than men. In place of performing more works, they used to occupy prominent place in the society and they used to take important part in collective life. The form of society was matriarchal. Women occupied special status in the society where there was pre-dominance of agriculture. But in Neolithic societies whose economy was based especially on domestication of animals possessed patriarchal system in which males occupied status.

Self-sufficiency of families and villages: Although division of labour based on sex had come into existence, but collective specialisation of industries had not taken place in Neolithic societies. Needs of each family such as food materials, pottery, cloth, tools, implements were fulfilled within the family group. In this sense each family was self-sufficient in matter of food material, pottery, cloth, tools and implements. The members of the family used to make these things by their labour.

Like families, Neolithic villages were also self-sufficient. The members of the village had to produce food materials, fodder, potteries, tools, implements woods, etc. themselves. For these purposes they were not dependent upon the members of the other villages. Thus, Neolithic villages were self-sufficient. These villages also lacked specialisation. The main reason behind these were lack of means of transportation in Neolithic societies. In absence of cart, women had to perform the troublesome task of carrying loads on their heads and shoulders. Thus, it was not easy to transport goods of one village to another village. Therefore, the people of one village were not known to the people of other villages for essential goods. But self-sufficiency cannot be taken as lack of reciprocal contact. Uniformity of Neolithic culture in the world and use of shells in Europe made available from Mediterranean sea bear testimony in this regard. But this fact is important here that such contact or exchange was not an essential part of economy of Neolithic man. Such type of contact or exchange was not able to affect the self-sufficiency of the village.

Social organisation: What was the power to systematise the social life in Neolithic period? It is very difficult to say. Perhaps unit of social organisation of Neolithic society was tribe. Each tribe had a totem. This totem was early ancestor of a tribe. When Neolithic villages began to change into chalcolithic town in Egypt, these were named after an animal or bird. In Neolithic villages, elephant and eagle were totem of tribes. Some scholars believe that the king had also come into existence in Neolithic period. In villages, one big house surrounded by some common houses has been discovered. This big house would have been palace of the king. But this cannot be said with certainty. It is also possible that this big house would have been Panchayat house of Neolithic villages.

Religion and magic: The religion of Neolithic society was characterised by mother worship. Mother power cult was prevalent among them. A number of images of women have been discovered from Neolithic villages. Perhaps they had belief that earth was like mother. As mother breast feeds her children, in the same way mother earth breast feeds her children human beings by yielding grains. Mother earth can be pleased by worship and gift. Perhaps they had also faith that mother earth could be controlled by sorcery and sympathetic magic. That is why Neolithic man used to make image of woman which represented mother earth.

In many regions emphasis was placed upon males in production process. Evidences in this regard are penis images obtained from Anatolia, Balcon country and England. From these countries penis images made of soil and stone have been discovered. The above opinion is also supported by an another fact. In early civilisations developed after Neolithic period, generally a fertility drama was staged. In this drama, marriage between a king and a queen was held. The king and queen were allowed ceremonial union of sexes.

Ceremonial union of sexes was taken as fertility symbol. The main character of this drama was corn-king. As seeds die or is buried in the process of production, in the same way the king had to die in this drama as a process of production. This drama was also related to the practice of human sacrifice prevalent among the Neolithic tribes of Western Asia and Mediterranean countries.

Art: Neolithic art was not like cave art. The Neolithic art is characterised by image making of women. During Neolithic period, human groups living in Egypt, Syria, Iran, South Eastern Europe and Mediterranean countries used to make varieties of women images from soil and stone.

Death rituals and megaliths: Neolithic human groups used to bury their deads either in grave or in house in which they lived. Pottery, tools and food materials were also placed in the grave along with dead body. They used to pay more attention on death ritual in comparison to Palaeolithic man. Perhaps they had faith that production of grain was related to deads. In Mediterranean countries the deads were buried in grave dug out in the house below earth. In North and East Europe megaliths or great stones were erected to pay respect to the dead. Such practice to pay respect to deads was prevalent in Scandinavia, Britain and Southern England. More manpower would have required for this practice. The earliest form of stone burial in Europe is called Dolmen or Table Rock. In this a plane stone is placed over several stone pillars. Thus, it resembled like a table: Dolmens were covered with heaps of soil. The heaps of soil is called Tumlus. In such type of stone chamber, bone remains were placed. Tumlus and Dolmen together are called Burrow. In Britain, generally only one stone pillar was erected. It is known as monolith or menhir or long stone. These are found in small or big size or plain or engraved with pictures. These are like same stone pillars as are erected today in the form of memorial stone. Neolithic man believed stay of soul in those stone pillars. Menhirs were erected in lines. In this condition they are called alignment. Those menhirs which have been placed in the centre encircling stone frequently are called cromlech.

Sorcery: Neolithic human groups had faith in sorcery. In Merimde small stone axes have been discovered which have a hole. This was perhaps worn as fetish in the neck. The small axes in the form of fetish used to protect them.

Fig. 12.8: Menhir of Neolithic Period

Knowledge and science: Knowledge and science of Neolithic man was better than Palaeolithic man. In this period, man was able to assemble knowledge from his experience and experiments of several centuries. He had acquired knowledge about chemistry of baking pottery, bio-chemistry of cooking and agrology of crops, plants and fruit bearing trees. He possessed some knowledge about anatomy of human body. Some evidences of bones reveals that they were jointed affer getting fractured. Agriculture has very close relationship with climate and seasons. Sun, moon and stars have provided knowledge about climate and seasons. Neolithic man had started his step in this regard. For example, the inhabitants of Egypt during the last phase of Neolithic period had discovered that Sirius planet rose in the sky at the same time when there was flood in river Nile. In due course people of Egypt began to realise that flood in river Nile was caused by the planet Sirius. Such type of experiences revealed that planets controlled activities of man on earth. This is the basic principle of Astrology and Astronomy. In the beginning of historical era such opinions were prevalent. Perhaps the seeds of such belief were sown during Neolithic period. Some scholars believe that at some places sequence of megaliths were decided as per the activities of planets. If Astrology and Astronomy were so developed in Neolithic period, the explanation about sun, moon and star would have also been prevalent in those days. But it is very difficult to prove these assumptions at present level.

Fig. 12.9: Megaliths Found From Stonehenge of England

Achievement of Neolithic Man

By the end of Neolithic period man was able to build nearly all pillars of civilisations. Fire, essential tools, implement, agriculture, domestication of animals, pottery, cloth and house had come into existence. These things are basic needs even for man today. Man had also discovered art in Magdalenian period. Leaving script, metal and state man had done every kind of inventions to live in civilised manner. Script, metal and state were invented in metal age. Economically, too, Neolithic revolution attained success. Man had brought the nature particularly land under his control to some extent by agriculture and domestication of animal. This was the foundation stone of progress of human civilisation.

13
Evolution of Chalcolithic Man and Culture

Demerits of Neolithic System

Before dealing with Chalcolithic culture, it will be the fitness of the thing to present here demerits of Neolithic system. It is true that man was dependent upon nature for his survival. He had started agriculture and domestication of animal in Neolithic period. The Neolithic economy was very much successful from the point of view of short period. But from the point of view of long period, Neolithic economy had two main demerits. Firstly, Neolithic economy was not successful in solving the problem of increasing population permanently. The only solution to this problem in that period was to look for new land for cultivation and new cattle for domestication. In the beginning this work was very simple. When the population of a village was increased, some families of the village established another village and grazing ground in the neighbourhood. But the expansion of land is limited. A time came when human groups began to face problem in finding out another agricultural land and grazing ground. In this period of scarcity of land, some group began to take away land and grazing ground of another group forcefully. But practice of snatching of land was not successful in solving the problem of increasing population. Secondly, Neolithic self-sufficiency of families and villages had also made them helpless during flood, starvation, storm, fires, epidemic, etc. Neolithic families and villages were self-sufficient. They did not possess reciprocal relation between families and villages. Neolithic families produced grains only to feed its family members. They neither helped another family nor expected help from other families. As a result of this, they used to became helpless during earthquake, flood, epidemic, starvation, fire, etc. when such natural calamities prevailed continuously for two or three years, they used to become extinct.

New Inventions

In order to solve these two main problems of Neolithic economy, it was essential to get more production from the same land and to establish reciprocal relation between families and villages. After Neolithic period, man attempted to get success in this direction by his several inventions. In the world history, the period between 5000 BC to 3000 BC was famous for inventions. All important inventions leaving present day scientific inventions were made in this period. According to British neo-evolutionist anthropologist V.G. Childe, sixteen important inventions were made in this period. These important inventions were as follows:

- Use of animal as weight carrier
- Sailing sheeted boat
- Wheeled cart
- Plough
- Artificial irrigation by canal
- Cultivation of fruits
- Brick making method
- Glazing method
- Wine making method
- Solar calendar
- Money
- Script
- Numeral notation
- Invention and use of bronze

Pre-historic archaeological anthropologists call this period as copper age. Along with copper stone tools were also made and used. That is why it is known as Chalcolithic period.

Copper, bronze and urban Revolution: Copper smiths had started using copper made plough. This practice was helpful in solving the problem of increasing population by yielding more foodgrains from land. That is why villages began to grow gradually in their size.

Although villages were big in size, but now they were not self-sufficient. The form of social organisation also started becoming gradually complex. Due to invention of wheeled carts and use of animal as weight carrier, villages were now not in isolation. But in spite of all such inventions, stone tools were also in use. Although now villages were bigger in size, but they had not taken the shape of towns and cities. In this age the use of stone tools and copper tools continued together. That is why this age is also known as Chalcolithic age and culture is known as Chalcolithic culture. At the end of copper age, i.e. 4000 BC, man also started a new experiment to solve the problem of food materials. He selected the land of fertile river valley for cultivation. He made canal for artificial irrigation of the lands situated in river valley. The building of dam and canal was not possible by the people of small villages. This activities forced human groups to organise themselves in the form of towns. At the same time, man was able to invent the method of producing bronze. In other words rise of urban civilisation, irrigation by artificial means of canal and use of bronze took place simultaneously. Here, we are going to deal only with the achievements of man made during Chalcolithic period.

Chalcolithic Villages: Before dealing with the description of Chalcolithic villages, it will be the fitness of the thing to present the description of place of rising of Chalcolithic culture. Chalcolithic culture originated in those vast parts of land whose spread was from Egypt, Mediterranean countries and India. This included Nile river valley, Asian countries, Asia minor; Syria, Palestine. Asyria, Babylonia, Iran, Afghanistan and North-Western India. Today these regions are dry, but before historic period these regions had more and more rainfall. Today major parts of these regions possessed mountains and deserts, but they also had green river valleys and mountain valleys. Neolithic villages had emerged in these regions. Archaeological remains of copper age, too, have been found from upper layers of those Neolithic villages.

Villages of Egypt: We have very little evidences to throw light on the pre-historic age of Indus valley, but with the help of evidences available from Iran, Babylonia, Asyria, Syria, Palestine, Egypt and Creat, main stages of development of copper age civilisations can be reconstructed. In Egypt the oldest layers of copper age are called as Badarian and Amratian. The life style of these inhabitants were like

Neolithic. They were acquainted with copper, but they did not know the method of making tools by melting it. They believed more in other world than this world. Their grave ground consisted of valuable tools and ornaments. They exchanged them with valuable stones of foreign countries. This intention gradually gave rise to pyramid construction in due course. In later culture called Garjian invention of making tools by melting copper had come into existence. In this age, the people of Egypt came in close contact with the people of Mesopotamia. The burial place of this age are bigger and beautiful. The remains available from burial places are valuable and artistic.

Villages of Western Asia and Iran: As referred to earlier that the first layer of Siyalak in Iran and other places in Mesopotamia has yielded remains of Neolithic culture. The second layer of Siyalak, Syria and Asyria has also yielded the remains of Neolithic culture. But some changes have taken place. Here, the export and import of courie, shells and valuable stones were increased. Houses were made up of unbaked bricks and furnace for baking pottery was in use. The use of copper was also started, but the method of making weapon and tool by melting and casting was not invented. Attempt was made to make a tool by crushing and beating the metal. In this period use of fetish was increased. Both women and men used to put fetish. Temples were built for Gods. The temple of Iridu town in Sumer perhaps belonged to this period. Archaeological anthropologists have termed this culture as Halafian. The Halafian culture has been named after a place Tell Halaf. This culture was contemporary to Badrian culture of Egypt. The next period, in which third layer of Sayalak and Al'Ubai culture of Mesopotamia and Syria came, has shown the use of copper tool made by melting and casting method. Along with the use of copper tools, stone tools were also in use. Potters had started using wheel for making pottery. Traders had started using money for exchanging things. In Sumer hand made potteries were also in use. New temples were built in place of old temples. Al'Ubai culture seems to be contemporary of Amratian culture of Egypt. Perhaps cultures similar to this were prevalent in Azilian country, Asia minor and North-Western India. In next period, the fourth layer of Sayalak, Tepgawra culture of Asyria and Jamdet Nasr of Sumer had emerged. These cultures seems contemporary to Garjian culture of Egypt. In this period, Chalcolithic villages began to change into small town and cities. The towns of Asyria were very small in this period. The inhabitants of these towns had started using baked bricks and molten copper. In Sayalak fourth and Sumer big towns had come into existence. The inhabitants of these towns had invented script. They were also using copper tools. These towns were much developed politically. About these towns we will discuss in next chapter. Let us discuss here those inventions of copper age which laid the foundation of urban civilisation.

Production of Copper and Use for Making Tool: Using copper for making tools and implements was a revolutionary inventions for human life. The use of copper was not so easy as was the use of stone. In order to make a tool from stone, the stone fragment was broken by any one method and then grinded. But for the use of copper more technical skill was required.

The copper as a matter was superior to stone. That is why its use became popular soon.

Properties of Copper: The following are the properties of copper:

- It is a flexible metal. It can not only be grinded like stone but also be moulded easily. Desired shape can be given to it by beating with a hammer. Its sheet can be prepared. The sheet can be cut into pieces for making tools and weapons. This property of copper had come into existence in Amratian culture of Egypt and second layer of Sayalak.
- In the tools of copper one finds hardness as well as sharpness like stone. The copper implements like stone implements also show stability. Baked earthen pots and stones cannto be jointed after getting fractured, but copper implements can easily be jointed by melting and casting method. Copper implements are not broken. A number of repair works can be done by beating with hammer and polish by rubbing. In copper one also finds flexibility like wet soil and hardness like stone. As pieces of wet soil can be jointed, in the same way pieces of copper can also be jointed. But copper also possesses some such properties which are not present in wet soil. For example, copper can be melted. It becomes not only viscous like wet soil, but also becomes liquid like water. If it is casted in its liquid form, it assumes the shape of the frame

after cooling in which it was casted in liquid state. On cooling it gets its solid state and become hard. After melting, it was possible to cast it in the form in which frames were available in number. Casted copper can be refined by beating and polishing. In Sayalak third and Garjian culture the human groups were acquainted with these qualities of copper. The places where copper was available in impure state, it can be obtained in pure form by scientific method. There are many such stone which yield copper when heated with charcoal. In Sayalak third and Al'Ubaid people were quite acquainted with this method of getting copper.

- In the valley of Dajala and Farat and other such places where stones were brought from outside and this process was costly. The tools and implements of copper were cheap. It was because one tool of copper was equivalent to many tools of stone. The tool of copper proved to be more fruitful at the time of war. The stone tools can get broken any time, but this was not so for copper tools. The copper can be made more harder by mixing it with tin and glass.

Inventions Related to Agriculture: Man had started getting skin, milk and meat from the animal in Neolithic period. Now he began to think that he could perform more and more work with the help of cattle. He used to perform those work by own labour and hands. The tilling of land was very difficult task. This work was performed by women till now. Chalcolithic man invented yoke. The yoke was used to make a couple of oxen to plough the land. When the plough was invented, it cannot be said. Early ploughs were made of wood. That is why one does not get any evidence related to them. It can be said with certainty that the plough was in use around 3000 years B.C. in Egypt, Mesopotamia and perhaps in India. Its invention would have taken place many centuries ago. In Egypt plough perhaps emerged from spade. When the handle of spade was increased both sides, it could assume the shape of plough. Images discovered from burials of Egypt bear testimony in this regard. With the invention of plough, agriculture came into existence in the form which continued before the advent of industrialisation. Due to this, agriculture and domestication of animal came more closer to each other. Now each peasant had to keep plough and oxen to till the land. Oxen began to yield cow dung as manure. The use of manure increased the yield of the crop. Another important invention related to agriculture was cultivation of fruits. Now man began to cultivate fruit bearing plants like grain yielding crops. The cultivation of fruit influenced food and social life of man.

Inventions Related to Transport: The use of oxen in pulling plough influenced the transport system of Chalcolithic man. Previously the work of carrying goods from one place to another was done by man himself. Perhaps women had to share this responsibility more than men. They also used sledge cart which were probably pulled by dogs. Seeing oxen pulling the plough, they thought that the work of pulling cart could be done by the oxen. In Mesopotamia, sledge cart were in use till 2600 B.C. Use of animal as load carrier was also done. The first animal to carry load was ass. The ass as load carrier was used in Egypt before 3000 B.C. In Syria and Mesopotamia ass was definitely in use as load carrier in the beginning of 3000 B.C. Horse rearing was started to get milk quite early. But horse riding, use of horse in cart pulling and as load carrier were perhaps not invented. But from Indus valley saddle of horse have been discovered in doubtful condition. It has been dated as 2500 B.C. By 3000 B.C. Sumerians had started using some lequid to carry chariot. This animal might be horse, ass or donkey. Some scholars suspect it as camel.

Bullock Cart: Most revolutionary invention in transport was wheel. One finds evidences of definite use of wheel in Halfian age. Two or four wheeled cart were in use in 3000 B.C. in Tepgawara. By 2000 B.C. such carts were used in Sindhu to Creat and by 1000 B.C., these carts were used in China to Switzerland. But the use of cart in Egypt came into existence by 1600 B.C.

Water Transport: By 3000 B.C. human groups has stared the use of water transport with the help of air. In Neolithic period, man had learnt to make small and big boats to sail in river, lakes, seas, etc. He learnt to use sailing sheet (Pal) in copper age. The picture of boat with sailing sheet have been discovered on the pottery of Garjian and Al'Ubaid. This is strong proof regarding the use of sailing sheet in boat by Chaleolithic man. Boats with sailing sheet were also in use in Egypt and Mediterranean countries abundantly by 3000 B.C. This was the first occasion when some physical power was used as driving power. The method of transport was proved to be too cheap than other methods of transport.

Pottery: The revolution in transport also brought impact upon the pottery. Man used to make pottery with his hands till the end of Neolithic period. When he used wheel in bullock cart, then he thought of making pottery by wheel. He invented potter's wheel and began to produce ceramics in large scale in short time. The invention of potter's wheel gave rise to pottery industry or ceramic industry.

Impact of New Inventions

1. Rise of Special class and Beginning of Inter-dependence

New inventions made during Chalcolithic period brought direct impact upon the society. Till now special classes were not created. The division of labour based on specialisation of work had not come into existence by Neolithic period. The Chalcolithic invention brought specialisation of labour. As a result, new classes emerged in the society. These special classes were classes of copper smiths, potters and carpenters. Copper smiths, potters and carpenters were the early specialists of the society.

Fig. 13.1: Images Showing Yoke Placed on Shoulder of Oxen of Pyramid Period

Fig. 13.2: Use of Spade in Pyramid Period

They created special classes in the Chalcolithic societies. The work of these classes were so complex that a farmer or a cattle rearer was quite unable to perform them. These classes were gradually detached from the production of food materials (agriculture) and domestication of animals. On the other hand farmers and domesticator of animals had to produce more to get benefit of services of these classes. As a result, interdependency of individual and village was badly influenced. Let us consider the example of invention of copper. A number of things such as furnace, frames, pincers, etc. were required. It was very complex for a common man to prepare these things and to acquire the knowledge of making, melting and casting of copper. Coming out copper from stone, melting of copper and casting of copper were not less than magic. All men were not able to perform these works. Thus, after magicians and priests, copper smiths occupied the second position of specialists in the society. They were expert of metallurgy. Their science was so complex that other persons were unable to learn it or they were unable to teach it to the other people. They taught it to their sons and favourite pupil only. They did not produce foodgrains. They also did not domesticate animals. They feed their stomach by their own science known as metallurgy. Peasants and cattle rearers had to produce more food and cloth to get copper tools and implements from them.

Fig. 13.3: Farmers of Pyramid Period Driving Oxen

Like coppersmiths, some people were engaged in digging mines and getting copper by smelting stones. These works were also not less complex. Only those individuals could succeed as mine cutters who possessed knowledge of recognising and breaking rocks and method of cutting mines. The chemical process of getting metal by melting the raw material was also hard. For this purpose, a high temperature furnace was required. Only few people could acquire its pure knowledge. These people could not get time spare to do agriculture and domesticate animals. Copper was also not available everywhere. It was available only in those mountaineous regions where there was no habitation of human groups. Tin was available at very fewest places. That is why the use of copper and bronze gave rise to trade of essential goods not of luxurious goods. As soon as a society needed the use of copper, it became dependent upon other groups.

After coppersmiths, another specialist class was potters. During Neolithic period, women used to make pots by their hands as per requirement of the family. But after invention of wheel in Chalcolithic period, potter's wheel was invented. This invention gave rise to production of beautiful ceramics on large scale. As a result, specialist class potters came into existence.

An another specialists class in the Chalcolithic society would have been carpenters. Demand of plough, wheel, wheel cart, boat, etc. would have created an another specialist class of carpenters.

Fig. 13.4: Assess Carrying Load of Pyramid Period

Encouragement of Settled Life

Revolutionary inventions of Chalcolithic age gave rise to cultivation and production of fruits. There is difference between production of foodgrains and production of fruits. Foodgrain yielding crops were cultivated and harvested each year. But fruit bearing plants when cultivated at a particular place, they start yielding fruits after 5 years or

6 years. The place of foodgrain cultivation can be changed each year, but the place of fruit cultivation cannot be changed. The fruit giving plants give fruits for a period of 80 to 90 years. The Chalcolithic man started planting the fruits of grapes, dates and olive. The cultivation of fruits forced Chalcolithic man to adopt settle way of life. The cultivation of grapes gave rise to the process of wine making. It may be possible that the man would have been making wine from oat earlier than grapes. But it is certain that by 3000 B.C. wine had occupied significant place in Sumerian culture.

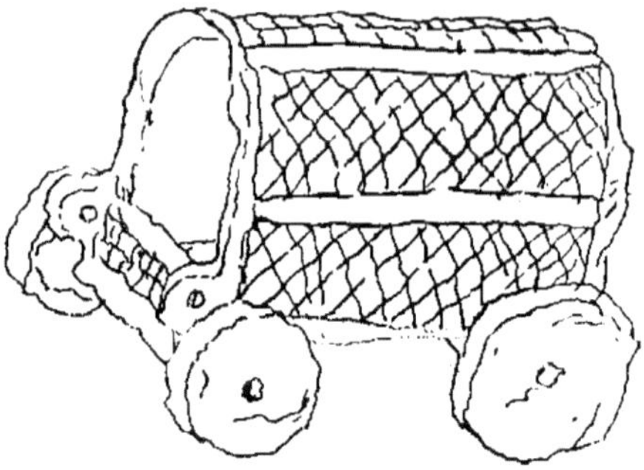

Fig. 13.5: Toy Cart

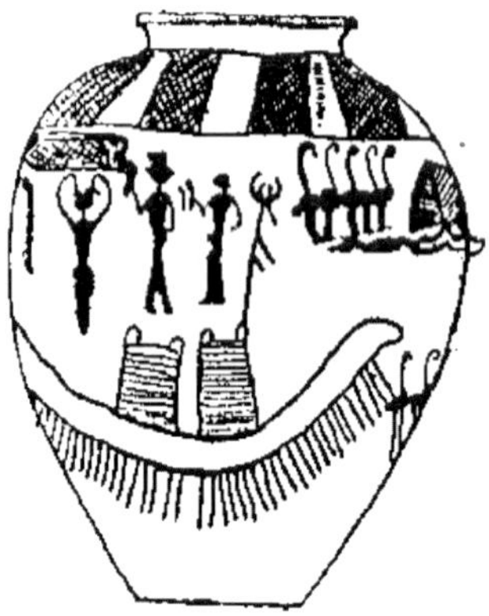

Fig. 13.6: Pottery of Garjian Period

Individual Property and Money

Inventions of new and new traits gave rise to individual property and money. The origin of money perhaps took place from amulets. Some totemic sign or religious sign were engraved on amulets. It was believed that the wearer of amulets possessed the mana or super human power of the sign engraved on that amulets. Gradually, this belief developed that when the sign of mana is imprinted on any object, the object is possessed by the power of Mana. In other words, the wearer of the amulet brought things under his control by imprinting the sign of his amulets worn. The power of amulets punished the culprit. From amulets emerged money in due course.

Fig. 13.7: Potters of Ancient Egypt Making Pottery on Wheel

Change in Social Organisation

The sense of owner could be shown not only on objects but also on man. During Chalcolithic period, different groups used to come often in conflict. These groups used to come under war frequently. To punish the defeated persons in war, slavery came into existence. In Other words man learnt to domesticate man through war. Another important change in society was related to position of women. During Neolithic period, the credit of inventing a number of traits had gone to women. The family was matriarchal and status of women was superior to man. But in Chalcolithic age, all inventions were made by men. The position of men became superior to women in this period. The new inventions of Chalcolithic period had freed women from the work of carrying load, making pots and doing agriculture. But their social status came down in this age. Now the form of society became patriarchal. The head of family became male. With headship, he also became owner of family tools, implements, property and slaves. All members of the family came under his control. They all began to obey the male head of the family.

Fig. 13.8: Two Individuals Carrying Pottery with the Help of Wood Stick Found from Khafza 3000 B.C.

Generally, in a group a person possessing more property and slaves used to become leader of the group in war. When he succeeded in war, he emerged more powerful. Now he occupied the post of head or *mukhia* of the tribe. His property was inherited by his son. Gradually, the post of head or *mukhia* became hereditary. The *mukhia* gradually became real king by staging agricultural drama.

14

Evolution of Bronze Age Man and Culture

By 4000 B.C. in the end phase of Chalcolithic age, man began to produce bronze and bronze made implements. There is no much difference between copper and bronze. Copper is flexible than stones. The edge of implements made up of copper can be destroyed easily due to its flexibility. But when some tin is added to copper, its hardness is increased. This tin mixed copper is called bronze. Its invention would have taken place suddenly. Tin would have been added suddenly at the time of melting copper and its hardness would have increased. The implements made up of this mixed metal would have proved more powerful. This observation would have made the man to know the value of bronze. It is very difficult to say when and where this invention took place for the first time. But it is certain that bronze was in use in Indus countries, Egypt, Creat and Sumer in 3000 B.C. Its use started in Troy in 2000 B.C. After 2000 B.C. the use of bronze began in rest of Europe. It is important to mention here that South India, Japan, North America and Australia possessed many parts where copper and bronze never came into existence. On that places, man enter into iron age directly from stone age.

Urban Revolution

Archaeological researches on bronze age proved about the emergence of urban revolution. Let us see as to why urban revolution came into existence in bronze age. There is no doubt that such inventions as production of copper and bronze, their tools and implements, plough, wheel, cart, boat with sailing sheet were full of revolutionary possibilities. But proper benefit from these inventions could not be taken without reorganisation of the society. The proof in this regard is tribes living in Syria, Iran and Mediterranean countries. The tribes of those countries were aware of copper, bronze and other inventions. But they could not make progress because they could not reorganise their society and their social system remained same as was before. The situations in the river valleys of Nile, Farat and Dazla and Indus were quite different. This great part of earth had started becoming dry from the period of Holocene. Man liked to settle in this part of earth because it was suitable for agriculture and fulfilment of other needs. Thus, we see that by 4000 B.C. the population of man began to increase in Egypt, Sumer and Indus country. At the same time big towns began to emerge. These towns were quite small in comparison to modern towns of London and New York, but were quite bigger than Chalcolithic villages. That is why neo-evolutionist V.G. Childe has termed this chapter of human civilisations as urban revolution.

Egypt is a small country surrounded by desert, mountain and sea. But the river Nile has built a very fertile land in its middle by bringing and depositing soil in flood. This land is made of different layers of fertile soil which is 30 feet in thickness. It is 750 miles in length and 10 to 30 feet in breath. In those ancient period this country was so fertile that it was not impossible to get three or four crops in a year.

Sumer is geographically South-eastern end of that fertile crescent which begins from Palestine on eastern coast of Mediterranean and ends in the valley of Faras in South-east through Syria and Asyria.

As Egypt was created by soils brought by the river Nile, in the same way Sumer was created by the soils brought by the rivers Dazla and Farat. Thus, as Egypt was boon of river Nile, in the same way Sumer was gift of rivers Dazla and Farat. The fertility of land of this country was world famous. The yield was 80 times than seed. Sometimes 100 times yield than seed was also not impossible. Fishes were available in plenty in rivers and lakes. Date trees were also found abundantly on land. Thus, Egypt and Sumer both were countries which attracted human groups. But in order to make them living worth it was necessary to perform hard work. In both these countries, rainfall was very low. It is true that these countries used to face the situation of flood, but soon after flood the land used to become hard. Therefore, it was impossible to get success in agriculture there without the artificial means of irrigation. It was also essential to control the water of flood. Summer had one trouble more. It was formed in recent years by flood waters of Dazla and Farat. So, it was not hard, rather viscous. It was also full of villow forests. Without making the land dry and clearing the forests of villow, it was impossible to do agriculture there. Thus, the fertility of land was in vain. The inhabitants of small villages of Chalcolithic period were unable to clear the villow forest, to control flood water by erecting dam and to make canal for artificial irrigation. For these works it was essential for human groups to form a bigger organisation for this purpose. Even after making dam and canal, it was essential to protect them. That is why, in Sumer and Egypt it became essential for human groups to organise in a bigger group. For this purpose different human groups began to reside at a particular place in very big group. Indus valley countries also had similar geographical conditions. So, urban civilisation also emerged in Indus countries at the same time.

Emergence of Towns in Sumer

By 4000 B.C. one finds the emergence of towns in Egypt, Indus countries and Sumer in place of Chalcolithic villages. On this transition period, Sumerian remains have given us strong proof. Iridu, Ur, Ereck, Lagash and Larsa towns of this country possessed similar sequence of stages. Let us consider here the proof of Ereck as an example. The earliest remain of this town belonged to Halfian and Al'Ubaid age. There is difference of 50 feet in the remains of Al'Ubaid and the beginning of historical age. Archaeological anthropologists divide them into two cultural age namely Uruck and Jemdet Nasr. During Uruck cultural period, Ereck village changed into town. Inana Devi temple erected in this period is 100 ft. long and 245 ft. wide. This period ended in 3500 B.C. This period is followed by another cultural period Jumdet Nasr. In this period, the glory of town increased. Precious stones from other countries began to be brought more and more. Tools and money began to be prepared by glazing method. Light chariots began to be made. Script and numerals were also invented. With the invention of script and numeral, it became easy for literature and scholars to compose their creation. It also became easy for traders to keep the record of things and price. The priests also started maintaining the register of income and expenditure in the temple and to perform magic. It also became easy for the king to preserve his achievements. Thus, by 3000 B.C. Sumer witnessed the end of pre-historic age and beginning of historic age.

Emergence of Central Power

The main basis of civil life in Sumer and other contemporary places was organisation of society in better way. The success of a town depends upon the collective participation of the citizens. The citizens of the towns had to perform such work as erection of temple, construction of dam, making canal and construction of buildings collectively. These works were common construction work of the citizens living in the town. For this purpose, it was essential to formulate a common plan of construction work. It was also essential to have sufficient means and manpower to get those works executed. It was also essential to have sufficient foodgrains and other things in the store to pay wages to labourers. Again, it was essential to have a central power to execute these plans systematically and to impose discipline upon the citizens of the towns. For all these works centralised power developed in towns.

The Priest Authority of Sumer

The responsibility of maintaining the town systematically was shared by the God of the main temple of the town theoretically and main priest of the town practically in Sumer. Here land was regarded as personal property of God. Temple was treated as Palace of the God. The priest commanded respect as representative or viceroy of the God. The chief priest used to maintain the system of the town with the help of other priests and by the order of the God. Every citizen was slave of God. He was bound to cooperate other citizens in construction works. Several classes of citizens emerged. These classes were of artisans, agriculturists, artists, servants and clerks. They began to work collectively under the discipline of the chief priest. Priests used to keep record of income and expenditure on earthen bands. This system continued in summer till the political integration of the country was not done. Under the leadership of Sargon first, political integration of the Sumer had taken place. There was also a central authority in Egypt to look after the system of town. In Indus Valley, too, some kind of powerful government was in existence. But the script of Indus has not been deciphered. Therefore, it is very difficult to say about the central authority. He may be a feudal lord, a king or a priest.

Foreign Trade

The farmers of Sumer, Egypt and Indus had to produce more and more foodgrains. It was because of emergence of those classes who did not participate in agricultural production directly. But it had one reason more. All these countries were such where all essential goods were not available. In Sumer, there was non-availability of stones and copper. Woods for building construction were also not available. Egypt was possessing stones, but lacking copper, wood, melchite, precious stones and resin. The inhabitants of Mohanjodaro and Harappa were lacking valuable metals. In short, it can be said that bronze age towns were not self-dependent like Neolithic and Chalcolithic villages. They had to import things for their need fulfilment. They had also to produce extra foodgrains. This fact was essential from the point of view of emergence of urban life.

In Sumer, the trade with foreigners was in the hands of member traders of temple. In Egypt, too, there was dependent existence of traders as class. But about Indus countries nothing could be said in absence of proof. It can be said with certainty that Indus countries had trade relation up to Sumer. The traders of these countries used to do export and import of foreign goods through merchants. In order to provide facility to those merchants trade centres were established here and there. The rulers of different countries felt the need of army to protect traders of their countries and to provide safety to the groups of merchants while they were travelling. Evidences about war by the rulers of the countries to protect the interest of traders have emerged by 3000 B.C. It had also become essential for them to settle reciprocal dispute between traders, merchants, artisans, agriculturists and other classes. For this purpose they had state workers, clerks. They also established law court to punish the culprit and determine the punishment. For courts, formulation of laws was essential. In the beginning attempts were made to punish the culprit as per existing traditions. But in course of time law codes were formulated to bring uniformity in customs of different places.

Fig. 14.1: Showing Brick Making Technology of Bronze Age

The temple priests had to maintain the record of property of traders and trade related matters. Thus, the script also emerged along with the emergence of towns. In the same way accountancy, numerals, Standard weights, measurements and geometric law came into existence. Also it became possible to record folk tales and other kinds of knowledge with the invention of script. By this creation of literatures their preservation took place. In order to help the agriculturists solar calendar was made by studying different planets. The invention of script also proved helpful in the progress of astrology and astronomy.

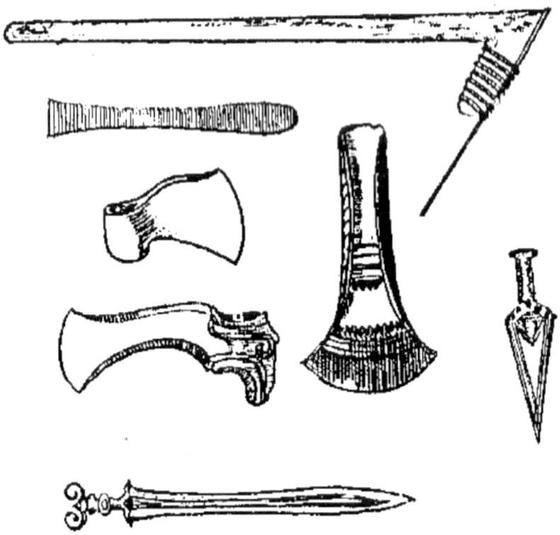

Fig. 14.2: Bronze Age Tools

In order to express his right on property and to make imprints on goods to be sent outside, seals were needed. This need gave rise to the invention of lapidary. Seal makers emerged as a class. This enhanced the demand of people having knowledge of glazing and artists of making glass.

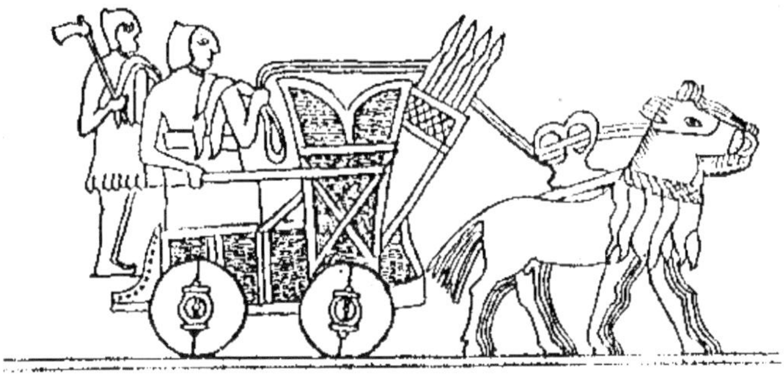

Fig. 14.3: Sumerian Chariot

In order to lead a settled life in town it became essential to erect building to lead a happy life. First of all he paid attention towards his buildings. He used to erect huts of villow during Neolithic and Chalcolithic period in Egypt and Mesopotamia. He invented bricks in Bronze age. Unbaked bricks were prepared with the help of frame. They were dried in sun. The inhabitants of Indus used fire baked bricks in plenty. The invention of bricks cleared the way of erecting building in place of huts. As a potter can make different types of pots, in the same way an artisan can make different types of bricks and a mason can joint them in different styles to erect a building. The early building of bricks resembled very much with huts. But gradually the size of building became bigger and bigger. The arch was invented in Sumer and Indus countries by 3000 B.C. By 3000 B.C. double storey building were erected in Indus countries.

Fig. 14.4: Arch (Mehrab) Found from Sumer

Civilised Society in the Beginning of Historical Age

From above account, it is quite clear that Bronze age urban revolution brought fundamental change in the life style of man. When in the early centuries of 3000 B.C. historical age emerged, at that time civilised man was inhabiting in towns situated in river valleys. These towns were quite bigger in size and population than Chalcolithic villages. The area of Mohenjodaro was more than 1 sq. mile. In the Ur town of Sumer buildings occupied 150 acres of land. This town had a population of 25,000. Evidences of 750 dead bodies have been discovered from state burial. Lagash of Sumer was comparatively a small town, but its population was not less than 19,000. This town consisted of buildings made of unbaked bricks, baked bricks and stone fragments. The towns of Indus countries were planned in such a way that they availed proper benefit of water supply and cleanliness. The water supply and drainage system of Indus countries was so improved that such system was not found in many towns of middle age Europe. The farmers of this countries used such plough and potters used such wheel in which no improvement took place till the occurrence of industrial revolution. The basis of economic life of these towns was foreign trade. The groups of merchants of these towns carried goods on animals, bullock carts and boats and brought goods from one towns to another. The beauty of these towns increased by the visit of traders

Fig. 14.5: Technology of Cutting Stones of Pyramid Period

from foreign countries. These traders used script numerals to maintain the record of income and expenditure. They used earthen bands with printed sign on their goods. The rulers of these country felt pride in erecting temple, building canal, palace and burial. They also felt glorious in expanding the territories of their country through wars. They engraved their achievement on earthen bands. The excavations of these towns have not yielded tools related to hunting, gathering and agriculture. But they have shown evidences of burials, canals, temples, palace, images, furniture, seals and inscriptions.

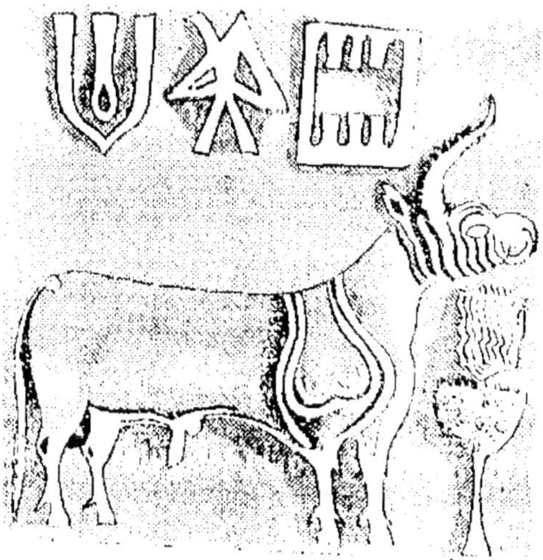

Fig. 14.6: Decoration of Houses in Bronze Age

In short, it can be concluded that all traits related to town life, happy life, civilised life and facility-ful life had emerged in copper and bronze age by 3000 B.C. In next 2000 years to come man could not enhance the inventions other than inventions of alphabets, iron and iron tools. That is why the age of bronze age urban revolution is termed as the age of birth of civilisation.

The picture of birth of civilisation presented above includes the urban life of Egypt, Sumer and Indus countries. But this cannot be meant that all the three civilisations were same. There was basic difference between civilisations of all these three places. Sumer and Egypt possessed political system entirely different. It is possible that Indus countries would have been possessing some other system. One

Date	*Egypt*	*Northern Mesopo-tamia*	*Sumer*	*Iran etc.*				*Indus*
		Sargoni Age			Hisar Third		Anno Third	Jhukar
2000? BC	Ancient State	Early Dynasty Age	Early Dynesty Period			Susa 2nd		Harappa
							Anno 2nd	Amari
3000? BC	Prag Dynasity Age	Ninweha Age	Jamdet Nasar	Siyalak Fourth	Hisar 2nd			
	Garjian Madian	Gawara	Post Uruk		Hisar Ist	Susa Ist	Anno Ist	
3900 BC			Early Uruk	Siyalak third				
	Amratian / Badarian / Mermad	Ubaid	Post Al' Ubaid					
4400 BC			Early Al Ubaid	Siyalak 2nd				
5000 BC	Tasian Fum	Halaf Hasua Jamo	Proto Al' Ubaid	Siyalak Ist				

Chart: Chronology of Chalcolithic and Bronze Cultures

also finds differences in other areas of life. Sumerian society was divided into a number of independent towns. The centre of collective activities of these towns were temples situated in the towns. In Egypt political integration had taken place quite early and power was in the hand of king in place of priests. We do not know much about the political system of Indus countries but it is certain that it had political system different from Egypt and Sumer. Such type of differences were also prevailing in different areas of life of Indus country and Sumer and Egypt. The earliest building of Egypt was state burial and the earliest building of Sumer was temple. We find the use of scripts at all three places, Indus, Egypt and Sumer, but the scripts of two places are not similar. In Egypt scripts were used for the first time on seals and in temples. But script was used in summer on earthen bands to maintain the record of income and expenditure of temple. Bronze was in use in all these three countries. But implements made by bronze smiths differ profoundly. Indus valley civilisation is different from Sumerian civilisation in town planning, images on seals, state burial, religion, dress and ornament, knowledge, science and living. These traits are also different in Sumerian and Egyptian civilisations.

Thus, it can be said that bronze age was not only the age of birth of civilisation, but it was also an age of birth of special civilisations. In this age man succeeded to give up his savage and barbarous life. He could succeed in giving birth to civilisation, the most evolved form of culture. But the form of civilisation became different in different regions due to adaptation with local environment.

15

Methods of Dating

In describing bio-cultural evolution, it will be the fitness of things to present here methods of dating and their relevance in bio-cultural evolution. Whenever, pre-historical materials are discovered while excavating a pre-historic site, a number of questions appear before a scholar interested in the study of pre-historic archaeological anthropology. These questions are about the age of pre-historic materials, geo-climatic condition of those times, plant and animal species living in those environment, builders of cultural materials and techniques involved in fashioning tools and artefacts. Pre-historic materials discovered during excavation of a pre-historic sites may include bones, teeth, antlers, ash, seeds, tools, artefacts, earthenware, etc. These materials found from the womb of earth tell us interesting story of bio-cultural evolution of pre-historic time.

The purpose behind dating pre-historic materials is two-fold: (i) To establish and study chronological arrangement of fossils and other evidences, and (ii) To identify and reconstruct evolutionary interrelationships and bio-cultural history of early man.

There are two chief methods by which dating of pre-historic archaeological materials are done. These are:

(i) Relative method of dating and
(ii) Absolute method of dating.

Relative Method of Dating

When the date of pre-historic materials is determined with the help of associated finds, it is known as relative method of dating. Here dating is not done in terms of actual years. Rather date is expressed in terms of older or newer by comparison between two samples or materials. Relative method of dating is also known as indirect method of dating because in this method dating is done indirectly by employing techniques of geo-chronology when archaeological materials are found in geological context. There are a number of methods by which relative dating is done. These methods are as follows:

- Typology or seriation
- Stratigraphy method
- Geological method
- Tropical rivers and lakes
- River terraces
- Palaeontology: Dating by using animal remains
- Palynology: Pollen analysis
- Astronomical method
- Chemical analysis of Collagen
- Chemical analysis of Fluorine
- Chemical analysis of Uranium
- Chemical analysis of Nitrogen
- Patination
- Rate of accumulation of cultural/natural deposits

Absolute Method of Dating

When the date of pre-historic materials is expressed in absolute or actual years, it is called as absolute method of dating. This is also known as direct method of dating because in this method associated finds are dated directly and date is determined in absolute years. There are a number of methods by which absolute dating of archaeological materials are done. These are as follows:

- C^{14} method
- Potassium-Argon or $K^{40} - A^{40}$ method
- Thermoluminescence method
- Archaeo-magnetic method
- Varve analysis method
- Dendrochronology or Tree-ring analysis method
- Fission track dating method

- Obsidian hydration method
- Amino-acid analysis method

Relative Method of Dating

Typology or Seriation: This method of dating is based on the principle that those pre-historic objects which reveal similarity in form and manufacturing technique belong to same age. In other words objects with similar form and manufacturing technique will be of contemporary period. Development and degeneration both may take place in manufacturing technique, but they follow a definite order. Thus, in determining the chronology of culture, typology plays significant role. Typology is also valuable for understanding technological revolution of mankind because it is chief base of study of tools available from earth surface. Study of technological development of ceramics is also based on the principle of typology. This method of dating was used for the first time by Petry. He studied pre-historic ceramics of Egypt by this method and determined relative age. Basically this method involves identifying minor stylistic changes in a given style or type through a period of time. Battleship can be observed, if frequencies of people using a specific style in a given time are represented as horizontal bar. The battleship reveals that there are very few people who start using style on their emergence. In the middle of ship its popularity increases and finally again its popularity decreases. If series constructed for a given type has some absolute dates, then any surface collection within the area can be ascribed a date by making comparisons with battleship designs.

The use of typology method has its own limitations. Limitations of this method are as follows:

1. It reveals only internal development of a specific period.
2. It becomes difficult to determine development or degeneration of culture of a specific age by this method. A number of manufacturing techniques of object have followed cyclic order. This problem is solved with the help of stratigraphy.

Stratigraphy Method: The principle of stratigraphy is a gift to us by geologist. Its importance was recognised first of all by them.

Thereafter, pre-historic archaeological anthropologists began to use this method for dating the archaeological remains discovered from excavation of pre-historic sites. According to principle of stratigraphy if no changes had occurred below the surface of earth, then pre-historic archaeological materials discovered from lower stratum would be older than archaeological material remains discovered from middle and upper strata. Remains discovered from middle stratum would also be older than remains of upper stratum.

This method seems easy but in practical it is not so easy. Researchers have to face a number of trouble in this method. The first trouble is that in all places and all environment deposition of strata is not similar. In caves, the speed of deposition is slow in comparison to other places. There is also difference in speed of deposition in hills, slope, plains and peats.

The second trouble is the great geological up-down brings changes in the strata. The study of stratigraphy of river banks and glacial regions are based mainly on survey method.

In river terraces, deposition of different periods are not found at same place.

Places of long period habitation sites reveal a different kind of stratigraphy.

Geological Method: There is very close relation between pre-historic archaeology and geology. The earth has preserved in its womb the evidences of man from his emergence on the earth to different kinds of environment in which he had lived his life. A number of changes have occurred in the surface of earth. Similarly changes have also occurred in climate, plant kingdom and animal kingdom. Geology tells us about these changes through geological timescale. When some archaeological materials are available from a definite geological layer, they provide us significant information from the point of view of dating. Generally it has been observed that stone tools are found in gravel deposits of river terraces and aeolin deposits of caves. They provide us significant information from point of view of dating. Geologically it has been established that Pleistocene is an epoch which forms a part of quaternary period which in turn is a part of Cainozoic era. In pre-historic archaeology Pleistocene epoch is the epoch in which man evolved on earth in the form of Australopithecus. The beginning of this epoch is believed to be anywhere between 5-4 million yeas from today. It is defined in terms of first appearance of some marker fossils. Marker fossils are known as Villafranchian flora and fauna ancestors of modern cow (BOS), elephant (Elephus) and horse (Equus) on land and Hyalinthecabaltica in sea. These evidences indicate that Pleistocene began 5-4 million years ago and continued around 8,500 B.C. To bio-cultural evolutionists, Pleistocene period is of great significance because just before its beginning Ramapithecus separated out to form a line of human evolution. Throughout this period man passed through at least four broad evolutionary stages and minimum of about twelve different cultural stages. One finds evidence of domestication of plants and animals only after the end of Pleistocene period. Quaternary period is divided into two: (i) Pleistocene, and (ii) Holocene. Holocene started nearby 12,000 years before today.

During Pleistocene epoch several glacial advances over Euro-Asia and America took place. All these period of glacial advances wee not of same intensity. There were also not intervening warm periods of same type. Scientists believe that during these glaciations compact ice of 2000 to 3000mt thickness stayed over the land for several hundred thousand years. The enormous amount of water trapped in the land born ice caused severe lowering of sea level exposing land bridge between many islands and their adjoining main lands. In many favourable coasts these ancient beaches have been found and named corresponding to the Alpine glacial order. The glaciation which descended down from north pole accompanied glaciers of all glaciated mountains gliding down to lower plains. In Europe, glaciers have been identified by giving them names of rivers on Alps. There are several synonymous for each of these names in different part of Europe, Asia and America, but to avoid complications only Alpine names are used here:

		Glacial	*Interglacial*	*Time*
PLEISTOCENE	*Upper*	Wurm		10,000 Yrs. B.P.
			Emien	
	Middle	Riss		150,000 Yrs. B.P.
			Hoxnian	
		Mindel		
	Lower	Gunj		
			Cromerian	400,000 Yrs. B.P.
			Villafranchian	4 Million Yrs. B.P.

The estimation of dates of 5 million to 500,000 years has been done through a radio active isotope disintegration technique known as Potassium-Argon (K^{40}—A^{40}) method of dating. Dating up to 90,000 years has been done by C^{14} method. The rest of date lying between 500,000 years to 90,000 years has been done through relative method of dating.

Now let us see how glacial and interglacial chronometer is applied in practice. Here a large number of plant types (described through identification of their pollens in soil) and animal types (described and defined through their skeletal remains in soil) specific to each

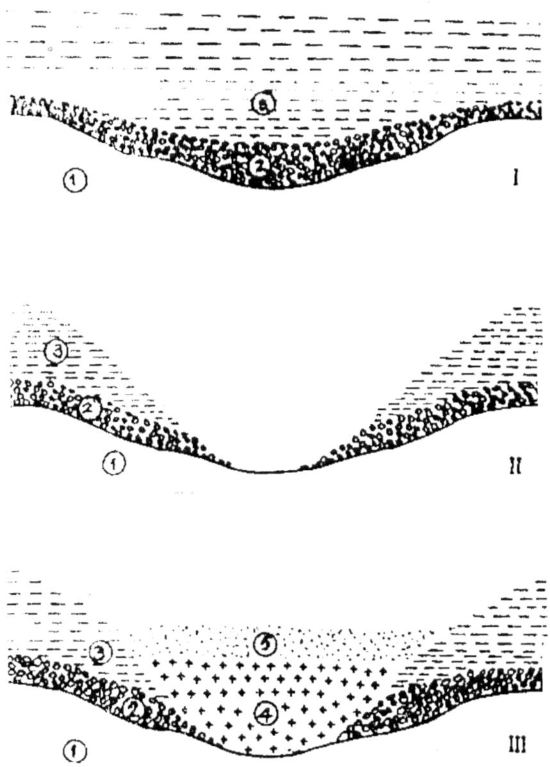

Fig. 15.1: Suggested Reconstruction of Peninsular India

Stage I: Aggradation (Interglacial Period)

Stage II: Erosion (Glacial Period)

Stage III: Aggradation (Interglacial Period)

of these climatic stages are established as guideline. In any pre-historic sites where these indicators of environment are found associated with cultural debris, the dating becomes possible by comparing the later with established guidelines. Archaeologically marker fossils are more fortunate because they are known to change into different species or sub-species at known periods within the Pleistocene.

Recently experts in Palaeomagnetic reversals recorded the actual size of reversal of North Pole into South Pole. In the last three million years, the North Pole of the earth seems to have become South Pole ten times. Each of these reversal has been named and dated.

Tropical Rivers and Lakes: It has been established that all rivers and lakes of the world are getting narrower and shallower. It is because tons and tons of debris have been accumulated in past. A series of stable periods have caused down cutting of bed and eventually changing the course. Deposits brought by rivers in past have been found sometimes more than one kilometre away from the modern flow of the river in a terraced structure. Studies of these available terraces in African and Asian rivers demonstrate that tropical regions of the planet underwent an excessive rainfall period during the time glaciations were occurring in the temperate regions. During these heavy rainfall periods rivers accumulated huge amount of debris. These were eventually discharged when water content dried up and could no longer carry them. Thus, a series of boulders and gravels are found in stratified deposits along the bank of ancient rivers and lakes. The periods of high rainfall are called pluviation. The dry periods between two pluviations are known as inter-pluvial. There are four such pluvial deposits found in East Africa. These pluvials are termed as Kageran, Kamasian, Kanjeran and Gamblian. Possibly these were occurring in the same periods when Europe was recording four glacials. Kageran took place during Gunz, Kamasian during Mindel, Kanjeran during Riss and Gamblian during Wurm. Some parts of Africa recorded two more wet phases after Pleistocene. They are known as Makalian and Nakuran.

River Terraces: Terraces represent the life history of a river. They are former flood plains formed along streams due to changing regime of river. These involve periods of aggradation or deposition and degradation or erosion. During the period of aggradation river will deposits silt and gravels and build up its bed. A number of world rivers present a succession of terraces. These terraces contain remains of animals and men who lived along the bank of rivers during formation

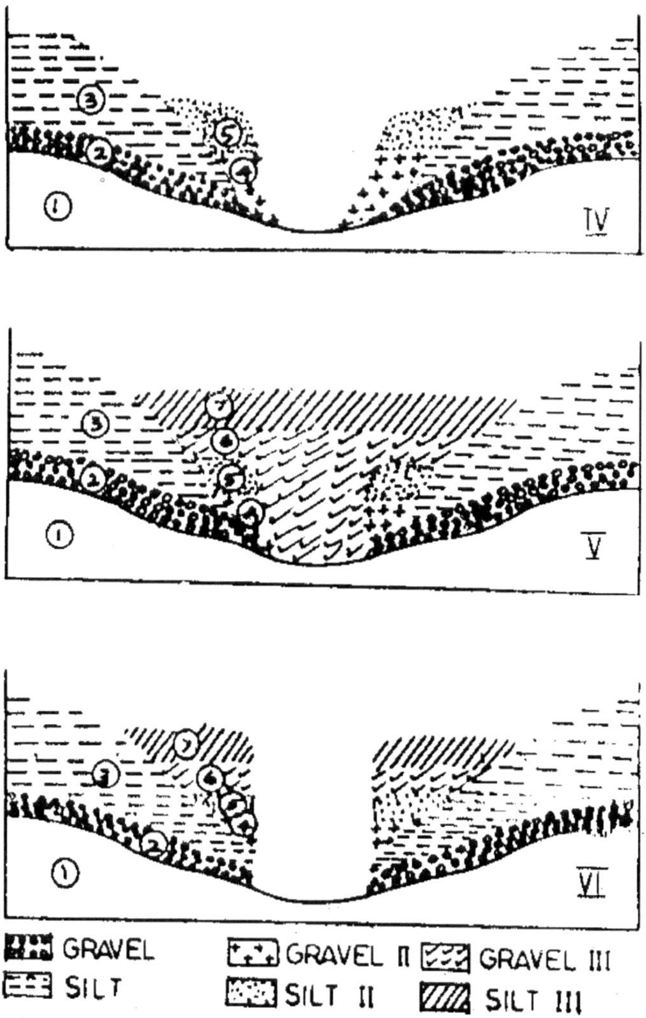

Fig. 15.2: Suggested Reconstructions of River Terraces in Peninsular India.

Stage IV: Erosion (Glacial Period)

Stage V: Aggradation (Interglacial Period)

Stage VI: Erosion (Glacial Period)

of their respective deposits. Study of these relics has an important bearing on chronology, climatic condition of early man and his culture. In case of river terraces, objects discovered from upper terrace will be older than middle or lower terraces. This is opposite of the principle of stratigraphy where objects discovered from lower layer are older than discovered from upper layer.

Palaeontology: Palaeontology is a science which deals with the study of fossils of animals and man. With the help of palaeontology relative dating is done. It is a established fact that bones are found in association with archaeological materials. The study of bones in association with cultural materials helps us to know what animals have lived with what cultures. Change in climate brings about occurrence of different animals and plants species. Change in environmental conditions reveals that certain species of animals have become extinct since appearance of man on earth. Evidences for Elephas antiquus (Forest elephant) reveals the existence of a temperate climate. Presence of E. Primigenins (a steppe elephant) indicates a steppe or tundra environment of almost glacial environment. Such relative methods of dating are useful for studying Pleistocene periods. The presence of forest and steppe varieties of reindeer reveals change of warm and cold periods during Wurm glacial periods in France.

The arrival of man in North America caused the extinction of mammals such as mammoth, horse, camel and several species of bisons.

Small species like rodents and birds, some molluscs and snails are very sensitive to climatic changes. Their presence or absence in archaeological sites reveals a change in climatic conditions. If these changes are related to varves, pollen or soil, it is easy to date them.

This method has some limitations. It is risky to interpret past climate on the basis of bones. It is because they are not random sample around the site. Rather they were brought by hunters mostly. Several animal species can live in wide range of climate. Man in past must have hunted animals around his living place. So he must have hunted only one species. Inferences about climatic tolerance of species that they have not changed over millennia is not a safe assumptions. Some animals prefer a particular environment, someother may be able to live in different environmental conditions. Preferences and tolerances are not well-known. Judgment on this aspect requires a great caution.

Palynology—Pollen Analysis: This is also a kind of relative method of dating. Palynology is a science which deals with the study of pollen grains of plant species. Pollens are not perishable. They are durable substance. They can be preserved for thousands and thousands of years. Study of plant species indicates changes in environmental conditions during the past. There is great diversity in plant species due to different kinds of environment prevalent in the world. Pollen grains of different plant species are different. They have been found buried in different geological strata. Analysis of evolutionary sequence of plant species from the study of pollen grains found from different geological strata is main theme of palynology. A level wise microscopic study of ancient pollen samples obtained from a vertical section of a pre-historic site helps us to trace the past vegetation history in relation to the prevalent climatic conditions and age of regions.

In the area of extreme flowering tree growth, pollen scatter is a common phenomenon. Under favourable climatic conditions pollen grains belonging to different plant species of different time period get deposited in the soil. Identification of plant species on the basis of pollen count enables us in making a pollen diagram. In pollen diagram relative frequencies of various species through time are plotted to show the changing vegetation. Peat can be dated by C^{-14+} method. Pollen analysis helps us to establish a very good chronology.

A Swedish Scientists Lennar Van Post developed this method of dating for the first time in 1916. In the beginning this method was confined only to the study of pollen from forest tree. But after two decades, it was extended to all flowering plants that scatter pollen grains. P.S. Martin, I. Gray and G.W. Dimbley in sixties of 20th century have sophisticated this method to the extent of recognising the climatic changes and chronology of terminal phase of Pleistocene and Holocene epoch.

It is also a fact that pollen grains are not preservable in all kinds of soil deposits. Researches done in Northern Europe revealed that Peat bag deposits are the ideal conditions for the preservation of ancient pollen. Other sources are dry sites, sands and clay. Acid soils with peat less than 5.5 may also preserve large amount of pollen. Recent researches have indicated that pollen analysis can be extended to dry environment and dry caves too.

Application of this method of dating requires a good knowledge of local plant, ecology and standard local sequences. By comparing the frequencies of pollen grains of living trees with those of past one can know various plant species distribution and their climatic conditions. A sample of pollen grains obtained from a vertical section of site can be described as pollen spectrum. Pollen spectrum tells us about plant history as well as climatic conditions of that area. Besides, we may get evidence from such levels which can be dated by C^{-14} method. By C^{14} method we can get absolute dates for various samples of pollen spectra of various levels. These pollen spectra with known dates can be examined. Cultural correlations and dates in number of years can also be examined.

Pollen analysis method has its own limitation. This method does not find favour with dry areas. Even samples from banks of streams cannot be dated unlike bogs because they are inorganic. Pollen frequencies might be disturbed by remains of domesticated plants or pollens brought by animals on their fur. When archaeological levels get mixed and overlapped, it will be very difficult to draw a correct picture by pollen analysis.

But in spite of all above mentioned limitations, it is very useful method of relative as well as absolute method of dating. If there is a worked out picture of climatic floral history of a region, it is easy to identify the pollen from an interglacial or glacial period without any date. But when prepared chronology is at hand, it is possible to make pollen profiles in terms of exact number of years.

Astronomical Method: While discussing geological method, chronological origin of glacial and interglacial ages has been described. Experts working with this change of climate came to the conclusion that planetary activities were responsible for climatic changes. Scholars were working on astronomical method since middle of 19th century. But in the year 1942 J. Blanchard propounded a very interesting theory in this regard. According to him, polar regions were transferred due to movement of earth, moon and solar gravitation. These astronomical activities were also responsible for chronological origins of glacials and interglacials. Since the time of transfer of polar regions can be calculated, therefore, calculation of dates of glacials and interglacials is possible.

So far as climatic change is concerned, it is not only due to transfer of polar regions. A number of complex natural processes are associated with it such as solar radiation curve and complexity of planet path. It has been established astronomically that earth moves around the sun. The motion of earth influences the intensity of solar radiation on particular surface areas of earth. Due to motion of earth solar radiation is changing not static. Experts working in this areas have attempted to measure different instabilities of solar radiation. On this basis, they have attempted to date them. The work of astronomist Mr. Melanchovich is praiseworthy in this regard. He established the chronology of solar radiation curves. The chronology of solar radiation curves was similar to the chronology of glacials and interglacials. Thus, there was a close resemblance between chronology of solar radiation curves and glacial ages.

Melanchovich calculated absolute date for solar radiation curves. The date established by him can be useful in establishing dates of geological ages and associated pre-historic archaeological cultures.

According to this astronomical method, the first phase of 1st Ice-age (Gunj) began 590,000 years ago.

From this date, the earliest fashioned tools by man came into existence. This human culture is known as Ipsvichian. Abbevillian and Clactonian culture came into existence during Gunj-Mindel interglacial. This culture existed till 60,000 years. During this period, the solar radiation was at climax. It had began nearly 500,000 years ago. The last phase of last ice-age Wurm began 25,000 yeas ago. Megadlinian culture evolved in this period. Thus, chronology of solar radiation established by Melanchovich shows close resemblance with chronology of ice-ages.

The dates established by solar radiation curves do not resemble with dates established by Radio carbon method. Different astronomists have different opinions regarding the dates of solar radiation established by Melanchovich.

Collagen Analysis Method: Collagen is a substance present in bones. In fact, it is fat and protein contents of bones. Collagen of bones does not disintegrate and disappear in certain situations particularly when bones lie buried in soil in a particular fashion in which there is no chance for the oxygen to enter into them. Under the influence of some microscopic organisms, collagen do not disintegrate for thousands of years. But as soon as bones undergo fossilisation, they start losing their collagen. There is relation between disintegration of collagen and fossilisation of bones. The rate of disintegration of collagen is directly proportional to the rate of fossilisation of bones. The older the bone, less is collagen content in it. Similarly, the younger the bones more is collagen content in it. Thus, by comparing the collagen contents of two bone samples, relative dating can be made.

Collagen from the bone samples are extracted by laboratory method. The method of laboratory extraction was revised in 1859 by Sinex and Faris. Collagen is also suitable for radio-carbon dating. K.P. Oakley was able to extract a small amount of collagen from the Piltdown man to the scientific world. By C^{-14} method, collagen present in Galley hill skeleton of middle Pleistocene has been dated as 3310 ± 100 years.

Fluorine Analysis Method: It is an established fact that buried bones contain fluorine chemical. The fact is that chemical composition of buried bones and teeth ungoes change by absorbing fluorine present in underground water and soil. The bones contain hydroxyapatite. The fluorine remains present in soluble fluoride in underground water. When bones are immersed in water fluorine ions remove hydroxyl ions from hydroxyapatite present in bones. As a result, fluorapatite is formed. The fluorapatite is insoluble in water and remains present in bones. The content of fluorine increases as bones become older. Bones get buried at the same time contain the same amount of fluorine. The older the bone, the higher is fluorine content and so more is age. The younger the bone, less is fluorine content and so less is age. The amount of fluorine can be determined by chemical analysis or x-ray crystallography.

Uranium Analysis Method: In uranium analysis method, analysis of uranium present in bones is done in laboratory. When bones of animals and men are buried in soil, they absorb uranium present in underground water and soil. As the time passes, uranium content of bone increases. The older the bone more is uranium content and the younger the bone less is uranium content. The analysis of uranium content is done by laboratory method or crystallographic method.

The bones contain hydroxyapatite. When hdroxyapatite of bones comes in contact with uranium ion present in water, calcium ion of hydraxyapatite present in bones is removed. The space is occupied by uranium. As uranium is a radio active, beta radiation is emitted due its radio active decays. The beta radiation per million per minute can be measured.

Nitrogen Analysis Method: Nitrogen analysis method of bones is also based on the same principle as that of uranium and fluorine. But fluorine and uranium both increase with the age of bones, but nitrogen contents decrease with passage of time. The younger the bone, more is nitrogen content and the older the bone, less is nitrogen content.

Nitrogen analysis of bone is done by laboratory method or crystallographic method.

There are a number of limitations in application of chemical analysis of bones. This method helps in dating bone specimen of same place or site only. It is not possible to compare bone specimens of different places because chemistry of absorption of chemicals by bone is not well understood as yet. There is also variation in the rate of absorption of fluorine by bones, teeth and antlers. Temperature and humidity also affect chemical action. Environments also differ in their chemical composition.

The goodness of chemical analysis of bones lies in establishing stratigraphy. No chemical analysis method is available for analysing absolute chronology.

Besides bones, teeth and antlers, analysis of shells and skins can also be useful.

Patination: Patination means chemical alterations of rock surfaces due to exposure of atmospheric condition. The amount of patina of a stone tool is useful at those sites which have stratified deposits. The stratified deposit sites reveal that tools of bottom level has more patina than the tools of upper levels. Thus, tools of lower levels are older because they have more patina. So, on the basis of amount of patina relative ages can be assigned to stone artefacts fashioned by man.

A.J.H. Goodwin had worked extensively on patination in 1960. He listed a number of variables involved in patina formation. He also recorded different types of patination.

Rate of Accumulation of Cultural/Natural Deposits: This method of dating gives us rough calculation of time on the basis of thickness of habitation deposits. In the beginning, this method was widely used by archaeologists because this was the only method of a dating available with the discipline of archaeology. Since the beginning of 20th century, this method was extensively applied by archaeologists working in USA and Near East. For example, R. Pampelly calculated that culture strata at ANAU accumulated at the rate of 2.5 feet per century, while in Egypt the rate of refuse accumulation ranged between 1.43 and 1.9 feet per century. Similarly, N.C. Nelson, W.E. Schench and E.N. Gifford determined the antiquity of refuse accumulation. They suggested a chronology of culture horizons of adjacent lower Sacramento valley. When their chronology was verified by C^{-14} method, it was found true to a considerable extent.

This method has been further applied in U.S.A. for dating natural glacial deposits, sedimentary varves, alluvial fillings and peats which overlie the cultural strata.

L.L. Loud and M.R. Harrington determined the dates of cultural deposits in lovelock and gypsum caves. Radio carbon dates also coincide with their dates approximately.

The problem associated with this method is that the rate of cultural deposits at any site is not constant. It depends upon factors like increase or decrease in population, use of several debris dumps, and lateral expansion of sites. M.L. Fowler had excavated the modock rock shelter in Illinois. This rock shelter contained a habitation deposit of 27 feet deep. A number of strata had yielded charcoal samples. Of them, eleven samples were dated by radio carbon method. Fowler was interested in calculating the rate of accumulation of cultural refuse. He examined the vertical position of each date samples with reference to quality of refuse between it and next upper dated level. For the period between 8000-5000 B.C., the rate of deposition was found to be constant at about one foot per 500 years. Between 5000 to 3600 B.C., it had increased to one foot per 300 years. Between 3600 to 3000 B.C., it has increased to 1.7 ft. per 100 years. Between 3300 to 2700 B.C., it decreased to one foot per 400 years. The variable rate of accumulation noticed at the above mentioned site created problem for the archaeologists as to why and how they had occurred in that order. Accordingly to Fowler, the variable rate would have taken place due to climatic conditions, number of occupants over time, variations in living patterns, etc.

Sir Mortimer Wheeler, Director, Archaeological Survey of India, had also applied this method for calculating the duration of time required for six successive building phases of the platform of Harappa citadel.

Absolute or Chronometric Method of Dating

C^{-14} Method: This method of absolute dating was developed for the first time in 1941 by Professor W.F. Libby of Chicago University.

This method is based on the fact that solar radiation strikes the upper atmosphere and converts a small amount of the atmospheric Nitrogen into radio active isotope called C^{-14}. Since all living organisms exchange gases from the atmosphere, C^{-14} reaches in their cells the same level as in the environment. When the organism dies, the trapped C^{-14} in cells begins to disintegrate back as Nitrogen. From the laboratory experiments, it has been established that half of any amount of C^{-14} gets disintegrated in 5730 years. In other words, half life time of C^{-14} has been established as 5730 years. By measuring the amount of C^{-14} left in pre-historic bone organic remains; one can calculate the time that has passed since its death. This method can give a reliable date up to 50,000 years. After 50,000 years, C^{-14} left is too little to be measured. C^{-14} method implies the measurement of C^{-14} atoms present in object and those that have disappeared from it.

All substances on earth—mass, liquid or gas—are made up of one or a combination of two or more substances which are called as elements. There are 92 such elements that occur naturally. For examples, gold, oxygen, carbon, etc. The smallest part of an element is known as an atom. Each atom has a nucleus. Around the nucleus thee are one or more moving bits called electrons. Atoms of different elements have different number of electrons ranging from one to ninety-two. For instance each gold atoms have 77 electrons. Inside the nucleus, there are one or more bits which are called as protons. The number of protons is always equal to the number of electrons moving around it. This is known as atomic number. Therefore, all atoms of a particular element have same number of protons and electrons. They all have the same atomic number and they all act exactly alike. The nucleus also contains another part called neutrons. Almost all atoms have neutrons. The atomic weight of an atom consists of the number of protons plus number of neutrons in the nucleus. The neutron number varies atom to atom of the same element, but all atoms of an element have the same number of electrons and protons. Thus, the atoms having the same atomic number (same number of electrons and protons) but different atomic weight (different number of neutrons) are called isotopes. C^{-14} is an isotope.

The isotopes differ in weights and behaviour. One is called stable and the other is known as unstable. The unstable isotopes emit radioactive rays and they change. For example hydrogen has three isotopes. The most common is the one with one proton and one electron. The second contains one electron, one proton and one neutron. The third consists of one electron, one proton and two neutron. The first and second are stable isotopes of carbon but the third one is unstable isotope of carbon. Therefore, third one is also radio active isotope. The unstable isotope changes into an isotope of another element. Most of the elements have unstable isotopes. The rays emitted by the unstable isotope are measured by the Scientists. Unstable isotopes are also called rare isotopes. C^{-14} is such a rare isotope of carbon. Carbon has three isotopes. C^{-12} and C^{-13} are stable isotopes of carbon. But C^{-14} is unstable isotope of carbon. That is why C^{+14} is radio active.

In the outer atmospheric layers of earth, solar rays attack continuously. When they get entered into atmosphere, neutrons of different energy are produced. Basically air is a mixture of oxygen and nitrogen. It has been proved experimentally that where there is inactive oxygen, a number of reactions take place in Nitrogen with the contact of neutrons. As a result of this reaction C^{-14} is produced. C^{-14} is a changed form of a special Nitrogen isotope. In the laboratory of sky, the newly produced C^{-14} atoms are oxidised in the radio carbon-dioxide. This is taken by plants along with other gases. There after, all animals get it by eating grasses. Non-vegetarian animals get it by eating other vegetarian animals. In this way all animals become radio active by taking C^{-14} in their body directly or indirectly.

In a common carbon, there are six protons and six neutrons. This is known as C^{-12}. C^{-12} is not radio active, but C^{-14} is radio active because it is an unstable isotope. C^{-14} gets mixed with oxygen of atmosphere along with C^{-12}. It has been estimated that the earth possesses nearly 80 tones of C^{-14}. Due to radiation, 20 ponds of C^{-14} are produced each year. Prof. W.F. Libby found out that all animals have equal amount of C^{-14}. Due to rotation of earth, the distribution of C^{-14} is equal and its absorption by all animals is also equal. All animals take C^{-14} and C^{-12} present in their body till they are alive. But as soon as an animal dies, the process of absorption of C^{-14} and C^{-12} also ends. After death, the process of disintegration of C^{-14} takes place. The process of disintegration of C^{-14} continues till the entire amount of C^{-14} is not disintegrated. The radio active atoms cannot be seen, but they can be located and counted with the help of an instrument.

Concept of Half Life: All radioactive isotopes have a concept of half life. According to this concept, after a certain time period, half of the amount of C^{-14} present at the time of death of an organism is disintegrated. C^{-14} disintegration occurs through beta radiations. The quantity of C^{-14} present in the sample is counted on the basis of beta radiations per minute per gram. For 5750 years, 7.5 beta radiations per minute per gram are produced.

There is no unanimity among the scholars regarding the half life time of C^{-14}. According to the count of Prof. W.F. Libby, the half life-time of C^{-14} is 5568 ±30 years. Some other scholars have calculated the half life time of C^{-14} as 5570 ± 30 years. Still some other scholars have calculated half life time of C^{-14} as 5000 years. According to half life time theory of C^{-14}, after the death of an organism, C^{-14} present in the boy is reduced to one-half in as much years, again in the same years it is reduced one-half of the half C^{-14}. This process goes on till the entire end of C^{-14}.

Suppose, the half-lie of C^{-14} is 5730 years, then in 11460 years 1/4th part, in 17190 years 1/8th part, in 22920 years 1/16th part are disintegrated. After 70,000 years no C^{-14} will be left in the bone of dead organism. On the basis of calculating the rate of disintegration of C^{-14}, dating is done. In this method, first of all C^{-14} present in ancient sample is calculated. Then C^{-14} present in recent sample is calculated. The difference between two kinds of sample is taken out. In the process of disintegration C^{-14} is changed into N^{-14}. It takes place through following formula:

$$- C^{-14} = B - + N^{-14} +$$

Technique of Calculation: Two types of technique have been employed for calculating the disintegration of C^{-14}. One is known as screen wall technique and the other is called gas proportional counting technique. The screen wall technique was used by Prof. W.F. Libby. In this technique, the fossil sample to be dated is first converted into carbon. It is then smeared on the inside of a specially built Geiger-Muller-Counter. By counting the beta radio active ions produced by it, its date is calculated. In this technique, there is a chance of sample under examination getting contaminated by the surrounding environment. Therefore, a better technique was devised.

Application of Method: In order to obtain a date by counting the Thermoluminescence glow of a ceramic pre-historic sample, first of all the sample is crushed into a fine powder. The fine powder is placed on a thin sheet of steel. The thin sheet of steel is put at a bottom of a box devoid of any light. The steel sheet is stretched between two connections to a source of electric power for a quick heating. Electric power is passed till powder reaches at very high temperature. The hot plate is covered by a round box containing a special liquid. This special liquid keeps out the heat waves produced by the thin steel plate. Above the round box a delicate instrument is placed for the purpose of measuring thermoluminescence glow emitted by the powdered ceramic sample. There is also an another instrument to measure and record the surface temperature of the material. The light and temperature are recorded in relationship to each other. As a result of this relationship thermoluminescence curve is obtained. The radio activity of the sample is also recorded. The age of the sample is determined by dividing the amount of thermoluminescence with average radio activity. This gives exact date in year since the object was fired last time.

Dr. Kennedy had tested this method on a number of terracotta objects of known dates. The potsherds from Greece dug up at ancient market place in the centre of Athens by Dr. Rodger Edward gave dates in absolute years as that of C^{-14} method.

Limitations: This method has its own limitations. Main limitations of this method are as follows:

- This method is useful only in case of those ceramic like materils which would have not been made impure after initial date of firing. In reality most ceramics have undergone chemical alterations. The chemical alteration of ceramics affects for thermo luminescent glow. Such an alteration cannot be distinguished from the glow produced by radiation. This is most significant limitation of this method.
- Substances which are found and cooled recently cannot be dated by this method. It is because radioactive elements require long period of time to give off rays to push electrons from their normal path to trap them in material and to send them back to their normal paths by high temperature.
- It is not possible to detect the glow with naked eye.

Advantages: This method has its own advantages, too. Its main advantages are as follows:

- This method gives an absolute date of ceramics.
- Ceramics are present in most pre-historic sites.
- It can date objects probably as old as 200,000 years B.C.
- This method is of very good use in detecting modern ceramic flakes from the ancient ones.

Archaeo-magnetic Method: In this method, absolute dating is done by studying the magnetism present in pre-historic objects. That is why this method is known as archaeo-magnetic method of dating. When this method is applied on geological materials or fossils, it is known as Palaeomagnetic method. Both archaeo-magnetic and palaeomagnetic methods are based on the principle that change in direction and intensity of earth magnetic field takes place from time to time. The earth keeps such evidences preserved. It is an established fact that earth behaves like a big magnet. The force of this big magnet operates all the time in, on or around the earth. The cause and origin of earth magnetism are still a mystery. It is also not known as to why it is not steady in its force or direction.

Near the Northern and Southern tips of the earth, magnetic lines of force plung down and sink deep into it. As a result South and North magnetic poles are formed. Besides, these magnetic fields, there are geographic north and south poles. These are two ends of the imaginary lines on which the earth turns. These were established by scientists and travellers century ago, while they were engaged in preparing the maps of water and land masses on earth. It has also been established that the points of magnetic north and south poles deviate from the directions of geographic north and south poles from time to time. Thus, the true magnetic north which at one time may be on one spot and years later might drift by hundred of miles to the east or west of previous spot.

Variations take place in the angle of declination between magnetic north and true north and in angle of dip of a magnetic needle. The angles of dip and declination vary from place to place and they have been recorded. These variations have been recorded for 400 years in London and for shorter periods in other cities like Boston, Rome and

Paris. These measurements indicate that the magnetic field of earth has changed in the direction at a particular rate.

A number of studies have shown that a record of pat angles of declination and dip is trapped in baked clay. The magnetic and haematite grains in the fire clay align with the magnetic field of earth surrounding them. When they cool, they become magnetic again such permanent alignment of rains is called as thermo remnaul-magnetism. This simply means the magnetism remaining in the object. The direction and strength of second magnetism are determined by the earth's magnetic fields in which cooling takes place.

It has now been established that geomagnetic north pole which guides compass points drifts constantly at the rate of about 600 miles for every hundred years. This magnetic direction follows a particular pattern which has been used back in the past. By finding out the path of earth magnetic field at the time when clay sample was fired and by comparing it with present magnetic direction, the absolute date of the sample can be calculated in years. The measurements are done in the laboratory. For this purpose, a specially devised equipment is used. Determination and calculation of geographic north and magnetic north are done in the field.

From the above account, it is clear that when magnetism enters into a rock containing iron ores, then its direction is determined by the current of magnetism of earth. Iron magnetic oxides are found in a number of rocks. These iron oxides are present in the particles of rock so long as they are not heated at a definite temperature. The definite temperature is known as curie point. Iron oxide particles of rocks attain motion above curie point. Then they attain blocking point. Therefore, below the blocking point, the mass of magnetism is maintained. All magnetic oxides do not have same curie point. Hematites attain curie point at 670°C. It is more than all iron oxides. Some oxides attain curie point at atmospheric temperature. Clay, bricks, furnace, earthen stones can be dated by this method. Burnt construction must possess the previous condition. They must not be transferred from one place to another. Baking place of movable objects be known within a radius of 1000 miles. The condition of objects at the time of baking be also made known. It is because in furnace, potteries are kept vertically but bricks are kept horizontally. Potteries and bricks samples must be baked properly. In practical furnace is most proper sample for dating by this method. The problems related to potteries is that they have a chance of remagnetisation at the time of cooking, breaking out of fire and performing death rituals. Bricks and tides also have problem of remagnetisation in breaking out of fire.

Limitations: Archaeo-magnetic method of absolute dating has its own limitation. The main limitations of this method are as follows:

- This method is limited to its applicability to regions for which accurate data of earth magnetic field and angle of dip are recorded for last several hundred years.
- Since such dates are not available beyond 1600 A.D., its applicability in pre-history is ruled out.
- The reliability of this method depends much on the availability of good records of change of magnetic path of near pre-historic sites within or about 1000 miles.
- The sharpness of dates depends upon a series of samples baked clay from a single area which have known dates against which one can measure clays without dates.
- The materials should be such which were not shifted form the place of original baking.

Advantages: This method has following advantages:

- Absolute date can be obtained form this method.
- Almost all pre-historic sites have ceramics and clay, therefore, this method of dating is useful in pre-historic anthropology.
- This method is best used for complementing dates obtained by C^{-14} method.

Varve analysis method: This method of dating was developed by a Swedish Baron Gerad D.E. Geer in 1878. But varves as annually deposited layers of silt was recognised by Heer in 1863. Varve analysis method is based on seasonal variations and climatic conditions of ancient time.

In Swedish, annual layers of sediments deposited at the bottom of lakes are called as Varves. In a varve two kinds of coupled layers are deposited respectively each year. For a layer of silt and other layer of clay or a layer of limestone and other layer of chert powder.

Thus, a varve is a two part core of sedimentary rock representing melting and freezing conditions of past. By counting the coupled layers of each year, the date of entire sedimentary deposit can be calculated.

The formation of varves takes place in those areas where glaciers discharge their water in stable water of lakes. During summer season melting of ice takes place in glaciers. As a result of which a huge amount of water is discharged in lakes and lakes have increased amount of water. Sands and clays brought by glaciers settle at the bottom of stable water of lakes. Heavy materials are settled first at the bottom in summer season. During winter season, water begins to freeze in glaciers and water does not flow in lakes. Lakes water also starts freezing. Light and micro particles settle in the surface during winter season. Thus, each year a pair of layers of deposit is formed. Thus, each year a narrow band is formed during winter and a wide band is formed during summer in the core. Each year this process of deposition is repeated.

Varve is formed of materials brought by melted water of glaciers. That is why in all conditions, it is related to end moraine. Due to background movements in end moraine, the layer of varve also move in backward direction in the same ratio. Thus, the study of varve will tell us as to how much years end-moraine would have remained at a particular place and how much years would have taken place in the formation of entire deposition layer.

Method of Calculating the Date of Varve: There are two methods by which the date of varve can be calculated. In one kind of method the relation of varves is established with known date of geological layers. In this method, the deposits of glacial and interglacial periods are helpful to a great extent. The climate of glacial and interglacial periods was dependent upon the extension and contraction of ice. Therefore, they can be correlated with end moraine and varve of the glaciers. In this way, varve can be correlated with any phenomenon of climate and pre-historic layer. Thus, dating can be done.

In another method of varve analysis, dating is done by the internal evidence of the layers. Peat and other organic sediments can be studied botanically and the climate when the layers were formed can also be established. Again this can be correlated with glacial and interglacial periods for the purpose of dating. To get rid of defects, both methods are employed. These methods appear simple in theory, but they are not so simple in practice.

Limitations: The main limitation of this method are as follows:

- Varves are formed only near the ice, therefore, varves are not found in those areas of the world where there is no ice. On this ground, most part of the world do not exhibit ice formation.
- In the places where ice was present during the Pleistocene, there was no evidence of sediments in the lake basin. This tells us about the recession of ice-sheets.
- There is problem to witness varve present outside Scandinavia till today.
- Even today, where varves are present, they do not exhibit a dates beyond 17,000 years.
- The melting of ice does not take place at uniform rates. Therefore, varves would have been deposited more or less frequently than annually.
- In spite of these limitations, the method has been applied successfully in Baltic area, North America, South America and Africa. But correlation of these sequences are not convincing.
- This method lacks precision. Dates derived by varve analysis are older than those estimated by C^{-14} method for the same event.
- This method has no direct utility in dating of pre-historic culture.

Dendrochronology or Tree-ring Analysis Method: Dendrochronology or tree-ring analysis method was developed by Charles Babbage in 1838 in USA. This method allows dating of wooden materials only. This method is based on the fact that each tree forms a ring each year. By counting the rings, the age of wooden objects can be calculated. The method implies that the trees growing in temperate zones exhibit clearly defined annual rings of growth. Annual rings of growth takes place due to effect of temperature, light and moisture. The width and spacing of rings vary seasonally. When the trunk of the tree is cut, such groups of rings become visible. These groups of rings have a common

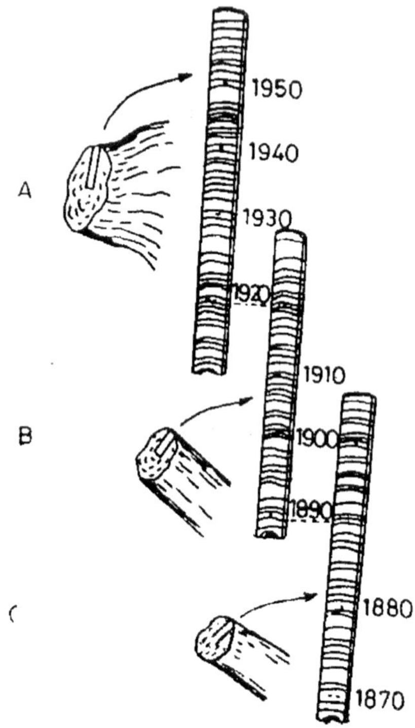

Fig. 15.3: Dendrochronology or Tree-Ring Analysis (for obtaining date by counting back from beams A to B to C).

A. Beam cut from a living tree

B. Beam from an old house for comparison of its rings with those of A.

C. Beam from an older house for comparison of its rings with those of B.

centre point. These rings indicate the sign of annual growth. Each ring differs from its near one in structure. It is because during spring season cells of bigger size and light colours are developed. But during summer, cells of smaller size and dark colour are developed. This speciality of formation of tree cells reveals clearly a difference between spring period rings and summer period ring. The cambium lying between the wood and the bark forms rings of two different kinds during yearly growth period of a tree. The thickness of tree-rings has direct relation with solar radiation. All trees of contemporary periods of a region exhibit similarity in tree-ring structures. The thin and thick size of rings of trunk of a tree shows similarity in sizes of rings of other trees. Dating by this method is basically dependent on this principle that all trees exhibiting similar tree-ring order belong to the same period. That is why principle of contemporisation is applicable here. If calculation is done from the rings of known date, the age of the tree can be determined. On this ground the calculation of past can be done. The counting of rings is done from the centre of the trunk to its periphery. The peripheral ring tell us as to when the tree was cut. This method can be employed for absolute as well as relative method of dating The calculation is done in the following ways:

Let us suppose that a tree was cut in 1990 AD. After counting the annual tree-rings, it was found that the tree was four hundred years old. It means that ring formation follows till 1590 A.D. (1990-400 = 1590). An another tree is also found which was cut in 1910. When rings of both trees will be compared, the 80 similar rings will be observed. If the second tree would have been of the same age, the ring order of past 320 years can be determined.

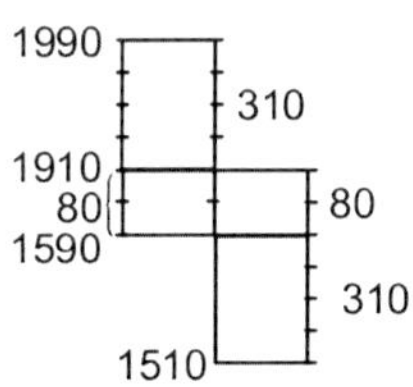

The regions which have dense forests, there one does not find any problem in studying the trunks of trees. In those regions, it is possible to establish a tree-rings order of hundreds and thousands years old trunks of trees. California has plenty of old trees, that is why scholars have been able to establish an order of tree-rings of 4000 years past.

Usefulness: The main usefulness of this method are as follows:

- It is useful in study of season change and solar radiation.
- Villages of pre-Columbia culture of South-west USA have been dated by this method.
- This method has been helpful in correcting the earlier dates obtained from C^{-14} method.

Limitations: The main weaknesses and limitations of this method are as follows:

- Synchronisation in tree-rings of trunks of tree belonging to different regions is not possible.
- All trees of a region also do not exhibit synchronisation in tree-rings.
- Woods were not used in different parts of the world in pre-historic periods. This method is not applicable for those regions.
- Wood and charcoal sample should be in plenty for the success of this method.
- The samples selected for study must exhibit clear tree-rings.
- Last ring on the log represent the time or terminal date when the log was cut. Such a date need not necessarily coincides with the time of its use.

Fission Track Dating Method: This method of dating is based on the principle that Uranium atoms decay by emitting alpha particles. This causes fission track damage on the surface of the material. Volcanic gas or some other minerals known to contain Uranium show such type of damages under microscope. For calculating the age of the sample, total uranium present in the sample and the density of tracks are counted. The ratio of two gives the date of the sample according to pre-determined constants for the rate of spontaneous decay.

Obsidian Hydration Method: This method is also known as water layer measurement method. Obsidian is a kind of stone made up of volcanic gases. It is black opaque glassy rock. It is also known as natural glass or volcanic glass. This glass was used for making tool in the past. Obsidian was cut and flaked to shape it into a tool. When obsidian is cut and flaked, the fractured or flaked obsidian absorbs water from the enviroment and the band of obsidian hydration layer is formed. Water enters into obsidian made tools at a known rate. Thus, obsidian made tool sample consists of water layer or dehydration layer. It is possible to calculate date of obsidian tool sample by measuring the depth of hydration layer. The rate of hydration formation can be determined in the laboratory under controlled condition. The thickness of hydration layer of the obsidian sample can be converted into years by using the rate of hydration.

Limitation: The application of this method is limited only to obsidian tool samples. This method is not useful for jasper, chert, quartzite, etc.

Amino-acid Analysis Method: Several amino-acids are present in the bones of human beings. When polarised light is thrown on them, some of them rotate the light to the left, while others rotate the light to right. The former type of amino acids are called as i-isomers and latter are called as d-isomers. Most of amino-acids found in living proteins are i-amino-acids. But when organism dies, i-amino-acids gradually change into d-amino-acids. This phenomenon is known as racemisation. It has been demonstrated that racemisation of a specific amino-acid called as aspartic acid has taken place in a period between 5000 to 100,000 years. Thus, by identifying the d-isomers of aspartic acid, date of bone samples can be calculated.

16

Evolution of Tool Techniques and Types

In order to supplement the knowledge of bio-cultural evolution, it is essential to deal here with evolution of pre-historic tool techniques and types. Pre-historic man had used different types of stones and bones for fashioning them into tools. He had employed different techniques of manufacturing tools in different pre-historic periods. As a result of this, different pre-historic periods reveal different types of tools. Different kinds of raw materials were also used in different kind of environment.

When man emerged on earth as a bipedal animal, his hands were now free to work. The only work for him in those early time was food collection for his survival and protection from wild animals. For food collection he used his hands to pull out bulbs of different plants, to break leaves, flowers, fruits and seeds of the plant and to collect forest produce. He was food gatherer but not hunter. Gradually, he learnt to catch small animals in the forest or nearby ponds and lakes. He began to catch small frogs, fishes, hares, rats, etc. with his hands. It is possible that he would have used wood sticks and some stone pieces for plucking fruits, kitting small animals and protecting himself from other wild animals. But as woods get rotten and stone pieces do not reveal human activities, therefore we do not have sufficient evidence. The only source of gathering and hunting activities are different types of stone tools and bone tools manufactured by man. The stone and bone tools have been discovered from the habitation sites of pre-historic man. The characteristic tool of lower Palaeolithic period was core tool. They were hand axes of various types. The characteristic feature of middle Palaeolithic period was flake tools called scrappers. The scrappers was of various types. The characteristic feature of upper Palaeolithic period was blade and bone tools. In fact blade was also a kind of flake. But all flakes were not blade. Thin and parallel flakes were called as blades. Blades were also of various kinds. The characteristic tool of Mesolithic period was bow, arrow and different kinds of microliths. Again characteristic feature of Neolithic period was Celt of various kinds, such as axes, adzes and chisels. Thus as man progressed from lower Palaeolithic to Neolithic periods, types of stone tools also followed evolutionary sequence.

Not only tool types followed evolutionary sequence, but techniques of tool production also revealed evolutionary sequence such as production of core tool, production of flake tool, production of blade tools, production of microlithic tools and production of polished and grinding tools.

Terminology

In the study of pre-historic tools and techniques, it is in the fitness of the things to present here the meaning of some pre-historic terminologies which are used frequently in the study of evolution of pre-historic cultures:

Artefact: An object of any kind produced artificially by man. It is smallest cultural unit of pre-historians. It includes core, flake, blade, bone, antler, horns, house, art, etc.

Industry: It includes all artefacts of same time discovered from a particular site. For example, core tool industry, flake tool industry, bone tool industry, microlithic industry, ceramic industry, etc.

Assemblage: This consists of all industries together found at one site.

Culture: A number of assemblages of same kind found at a number of sites. For examples, Palaeolithic culture, Mesolithic culture, Neolithic culture, Chalcolithic culture, etc.

Tradition: It refers to a particular segment of one and the same period. For example pebble tool tradition, hand axe cleaner tradition, Aurignacian tradition, Solutrean tradition, etc.

Phase: It represents a cultural chronological sub-period of a period.

Period: It contains more than one phase and represents a larger unit in terms of antiquity and cultural components.

Site: It is a place of different sizes where artefacts are found. It can be divided into different types based on artefacts found such as Palaeolithic, Mesolithic, Neolithic, Chalcolithic, etc. by location such as cave site, lake site, by type of activity such as kill site or camp site, by context such as stratified, surface, etc.

Core: It is starting stone material from which flakes are taken out. For example, pebble core, flake core, blade core, etc.

Flake: It represents a part of core with a flake surface with or without bulb of percussion and a platform. Flakes may be side flake, end flake, interminate flake, etc.

Chips: These are shapeless waste flakes resulted during the working of cores or artefacts. They are small and irregular in shape. They exhibit no sign of use or retouch.

Platform: It is a point at the end of side of a flake whose blow is given while removing it from the core. The platform is called faceted platform when it is prepared by removing small chips. The platform is called unfaceted when no previous preparations is made for the purpose.

Striking Platform: The surface on which a hammer hits a stone is called striking platform.

Point of Impact: The point on the surface where hammer hits is called point of impact.

Positive Bulb of Percussion: When a force enters a stone it imprints an elevated bulb on the inner surface of the flake right under the point where the hammer hits it. This is called positive bulb of percussion. The flake has positive bulb of percussion.

Negative Bulb of Percussion: The flake scar on the core maintains a deep point corresponding to positive bulb. It is called negative bulb of percussion.

Bulbar Scar/Flake Scar of Detachment: The surface on which positive bulb appears is called bulbar scar or flake scar of detachment.

Flake Scar: The corresponding area on the core is called flake scar.

Original Cortex: The surface of stone which bears no evidence of workmanship is called original cortex.

Blade: It is a thin flake with longer parallel sides. All flakes are not blade, but all blades are flakes.

Raw Materials

Pre-historic man had selected a number of rocks for making tool. The main rocks used in tool making are igneous rocks, sedimentary rocks, metamorphic rocks and minerals. The igneous rocks consist of basalt, diorite, dolerite, epidiorite, amphibolites and acid rocks.

The sedimentary rocks include sand stone and flint.

The metamorphic rocks are rocks of silica groups. They are divisible into crystalline rocks and crypto crystalline rocks. Crystalline rocks include rocks like quartz and rock crystal. Crypto crystalline rocks are chert, chalcedony, jasper, agate, cornelian, opal, obsidian, etc. Other metamorphic rocks are quartzite, schists, gneisses, etc. Crypto crystalline rocks are also known as mineral rocks.

All rocks were not used by pre-historic man of one and same cultural period. Man of different cultural periods had used different kinds of rock for fashioning tool. This was also related to the availability of rocks in local environment of habitation site of pre-historic man.

Lower Palaeolithic man used easily breakable rocks of raw material for manufacturing tools. The raw materials used by lower Palaeolithic man were obsidian, quartzite and metamorphic type according to availability nearby the habitation sites. In Europe lower Palaeolithic man had used sedimentary rocks like flint. In India quartzite and basalt were used as raw materials for making tools by lower Palaeolithic man.

Middle Palaeolithic man had changed his raw materials for making tool. He had also changed his tool. Flake tools like scrappers, points and borers were produced in this cultural period. Siliceous rocks of crypto crystalline variety such as agate, jasper and chalcedony obtained in the form of river pebbles or from the ourterops whose conchoidal fracture is more regular than that of quartzite. Quartzite of fined-grained variety was also used by man belonging to this period.

The Upper Palaeolithic man had produced his tools of blade and burin culture from coarse, medium to fine-grained quartzite of various shades besides some siliceous rocks.

Ice-age had ended with the end of Pleistocene. Palaeolithic way of life also came to an end. Now Holocene period began. Environmental condition became normal. Now it was congenial for man of Mesolithic period to produce technically advanced tool by employing the technique of producing microliths and composite tools. Metamorphic rocks of crypto crystalline and crystalline types of silica groups were largely used by the people of this period for producing geometric and non-geometric microliths.

Settled life came into existence during Neolithic period. In this period two kinds of stone industries emerged. One was pecked and ground stone industry and the other was blade and microlithic industry. The tools of first industries were made on igneous rock of basic as well as acidic types. The basic rocks such as basalt, dolerite, diorite, etc. were used frequently for making different tools of Celt family like axes, adzes, chisel, etc. Non-edge tools or domestic implements like rubber, mullers, querns and hammerstone were produced from acidic rocks like granite, granodiorite, epidiote granite. The choice of manufacturing tools was also dependent upon the locally available stones as raw materials.

During Chalcolithic and bronze age, the tradition of producing ground stone tools remained continued.

Evolution of Stone Tool Techniques

The chief techniques employed in manufacturing tools by man belonging to different pre-historic cultures were percussion flaking, pressure flaking, striking blades from prepared core and grinding and polishing.

Lower Palaeolithic tools were made by using block on block or anvil technique, direct percussion or stone hammer technique, cylinder hammer and hollow hammer technique. In addition to these techniques Clactonian, Bevalloisean and Mousterian techniques were also employed by the pre-historic man of Europe in lower Palaeolithic period. But these techniques were used in middle Palaeolithic man in Africa. In middle Palaeolithic period, technique of flake production which appeared in lower Palaeolithic period also continued. The technique of producing blade appeared in upper Palaeolithic. This technique continued in Mesolithic, Neolithic and Chalcolithic period, too. The technique of polishing and grinding was invented in Neolithic period.

Now let us discuss about techniques of stone tool production one by one.

Block-on-Block or Anvil Technique: This is a simple technique of manufacturing stone tools. This technique was employed for the first time by lower Palaeolithic man. In this technique the pebble or block of stone to be converted into a tool is struck against the projecting point of a large fixed block of stone called as Anvil. As a result of this, large and massive flakes come out with prominent bulbs of percussion. The bulb of percussion remains at an angle to the platform of the flake. The negative flake scars left on pebble are deep and vertical to the striking plane. They are much longer than wider. No working or retouch is possible on the artefacts prepared by this technique. This technique is convenient only for producing large and crude tool.

Fig. 16.1: Block on Block or Anvil Technique

This technique has a risk too. In this technique there was no control over fracture and purpose was defeated.

Direct Percussion or Stone Hammer Technique: This was most common technique used by pre-historic man for manufacturing his tools anywhere in the world at any time. In this technique two blocks of stone of suitable size were selected. Of two blocks of stone, one block acted as hammer and the other was fashioned into a tool. The hammer stone was kept in one hand and the other stone to be converted into a tool was held in another hand. Now, the hammer stone was struck against the other stone to be made into a tool. In this way flakes were removed with the help of hammer stone. This technique was related to direct percussion with the help of hammer stone, that is why this technique is known as direct percussion technique or stone hammer technique. Here, the shape, size and nature of the flake is determined by the direction and amount of force with which hammer was driven. The flake bears elevated bulb of percussion (positive bulb of percussion) and the core exhibits the corresponding depression (negative bulb of percussion). In this technique hammer usually hits the border of the core from the under surface not from the upper surface. The flake comes out from the under surface. Alternate flaking on both surfaces is nothing but the bifacial flaking and the flake scars from either surface intersect forming a zig-zag cutting edge. This technique is sometimes referred to as free-flaking technique. The block to be made into a tool is sometimes held on knee or on the trunk of a tree to serve as suitable pads.

Fig. 16.2: Stone Hammer Technique

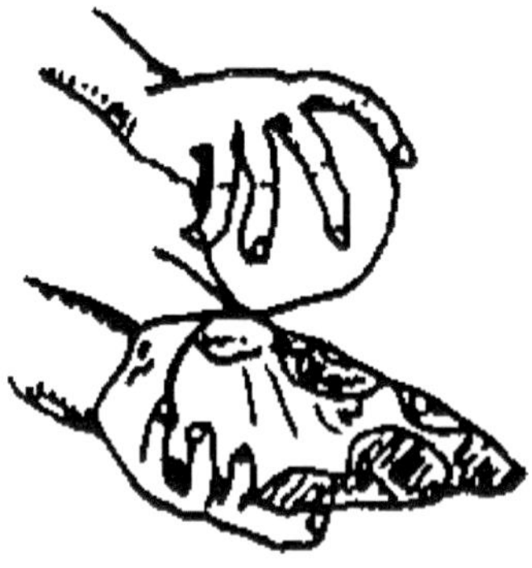

Fig. 16.3: Stone Hammer Technique

Step or Resolved Flaking Technique: In this technique the flake scars produced are shaped like a step, that is why this technique is known as step or resolved flaking technique. In this technique, the maker has to exercise control over the force of his stroke depending on the type of the tool he is making. In such situation, the force is driven towards the centre or thickest area of the module. The force becomes restricted and cannot spread over entire thickness. The force becomes exhausted halfway through resulting in the formation of a crack up to the distance the force enters. It is a kind of secondary flaking which helps in surface chiselling. The chiselling effect enables cutting the borders without sacrificing the thickness. As step like structure results at the face of the crack, so the technique is known as step flaking. Flakes removed by this technique are necessarily small hinge fracture flakes.

Cylinder Hammer or Hollow Hammer Technique: Tools with unusually shallow and elongated flake scars were discovered in pre-historic debris. After experimenting of many kinds of hammers, Dr. Leaky declared that such flaking could only be possible by the use of hollow bone or antler or wooden hammer. That is why the name of this technique was given as cylinder hammer or hollow hammer technique.

Fig. 16.4: Cylinder Hammer Technique

Tools produced by this technique reveal not only great technical development but also foresight in planning the design of tool. Such design of tool was not observed earlier.

The greatest advantage of using these organic hammers is their property of absorbing the reaction of force. This eliminate totally the shattering effect of impact. The bulb produced in this kind of scars is diffused. The flake scars have a more or less parallel running boundary ridge.

This is also a kind of controlled flaking technique.

Acheulian hand axes with shallow flake scars, even surfaces with biconvex or lenticular cross section and regular shape were prepared by this technique. Such hand axes were noticed first time at a French site named Saint Acheal in Somme Valley.

This technique was employed for making pebble tools and tools of Abbevillian and Acheulian hand axe industries. These were lower Palaeolithic tools.

Punching Technique: This technique involves the use of an intermediate puncher. The puncher receives the bow at its top end and transfer it to the core through the other end. This technique has the advantage of controlling both the magnitude and direction of force.

The puncher is held in one hand and the blow is delivered by the other. This technique was used in lower Palaeolithic, middle Palaeolithic, Mesolithic and Neolithic periods. The method is quite useful in knocking of undesired corners or protuberances on a rock surface without the risk of undesired damage to the tool.

Clactonian Technique: This is the oldest flaking technique used by British lower Palaeolithic man. Tools prepared by this technique were first found at classic site of Clacton-on-sea in Essex country, England. The tools are made of large massive flakes of this type or the flakes themselves have been used without any further working. These flakes are known from all over Europe during various stages of Palaeolithic culture. These flakes are typical. They are characterised by high flake angle and absence of secondary retouching.

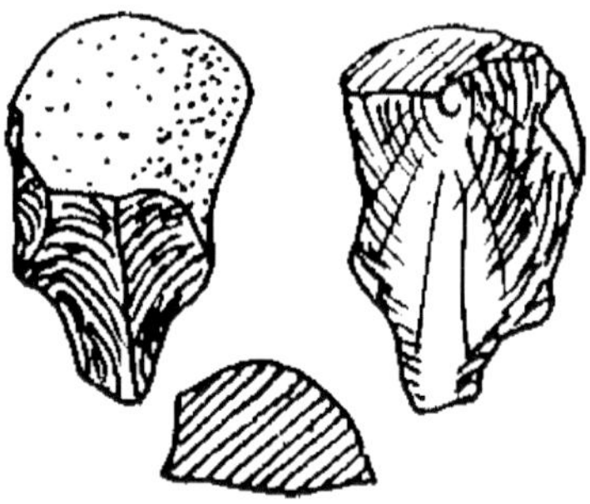

Fig. 16.5: Clactonian Technique

In this technique, a nodule with fairly regular surface is selected and a good sized flake is removed from one of its surfaces. Flaking is done with the help of a hammer stone. On observation a detached flake shows a small circular raised projection on its under surface. This raised projection is known as positive bulb of percussion. The flat surface which is hit upon is termed as striking platform. Flakes of this type can be removed either by direct hammer or by block on block or anvil technique. The Clactonian flakes bear the following characteristics:

- The flakes as a rule are large and massive and bear prominent bulb of percussion with ripples of force marked occasionally.
- The striking platform forms an angle of 100°-120° with flake surface.

- The striking platform as a matter of fact remains unflaked and original surface. Such unworked platform is known as unfaceted platform.

It is a common experience that to detach a flake from a pebble one has to give a blow in an inclined manner. The angle of this direction of force with the pebble surface is always less than 90° for obtaining a good size flake. For removing the second flake from the same surface, one requires increasing the angle of the blow. If this process is repeated, a stage comes when the blow delivered is at right angle and the core shatters in several pieces.

Critical Angle of Percussion: The Clactonian technique eliminates this limitation of simple percussion technique. Here, once a flake is removed from one surface, the flaked surface is used as striking platform to remove another fresh flake from the opposite surface. As a result, the flake angle of each Clactonian flake comes out as more than 90°.

The Clactonian technique is also known as swinging blow-blow-technique. It is because the pebble to be converted into a tool is hit against a hard block of rock or anvil with a swinging blow resulting in the removal of a flake. In India, middle Palaeolithic man had used this technique of manufacturing tool, but in Europe this technique was in use since closing stages of lower Palaeolithic period.

Levalloisian Technique: This technique is another advance and skilful technique of removing flakes. The use of this technique in manufacturing of stone tool appeared for the first time at the site of Levallois Perret, a suburb of Paris, France. The most significant feature of Levalloisian flakes is that they are prepared in advance on the core. The process starts with the rough trimming of the sides of core. Then cortex from the upper and the under surface is removed. This is done to facilitate flake scars to meet in the centre. All irregularities and uneven portions of the surface are also removed. This is done to provide rounded or semi-rounded shape to the core. This is also known as tortoise core.

In the second stage, a striking platform is prepared on the core. It is done by removing very small flakes on the core with a small punch or chisel along the margin. Preferably it is done at the shorter end or in the middle of the longer side where the two surfaces of the core meet and are perpendicular to longer or shorter axis.

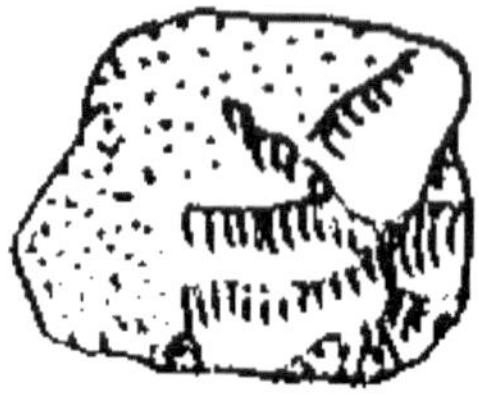

Fig. 16.6: Different Stages of Levalloisian Technique

Fig. 16.7: Different Stages of Levalloisian Technique

Fig. 16.8: Different Stages of Levalloisian Technique

Fig. 16.9: Different Stages of Levalloisian Technique

In third stage, flake is removed from the core. It is done with the help of a blow given directly by a sharp tool or narrow pointed tool known as Punch. The blow is given on platform. The core is held in left hand. The blow is at right angles to the platform or the axis of the core. The Levalloisian flake is roughly triangular or oval in outline. It has a clean under surface and a part of prepared platform. The under surface makes an angle of 80-90 with striking platform Levalloisian flakes have sharp edges and do not require any further retouch. This technique is also known as tortoise core technique or prepared core technique.

In Europe, this technique appeared towards the end of lower Palaeolithic period, but in Africa and India this technique appeared in middle Palaeolithic period. Acheulian hand axes were produced by this technique.

Difference between Clactonian and Levalloisian Techniques: The following are differences between Levalloisian and Clactonian techniques:

- In Levalloisian technique flakes are produced from a prepared core, but in Clactonian technique flakes are not produced from a prepared core.
- In Levalloisian technique, only one flake is produced, whereas in Clactonian technique more flakes can be produced.
- Levalloisian technique is economical, but Clactonian technique is not economical.
- Flakes produced by Levalloisian techniques are of quality and size but flakes produced by Clactonian technique are not of quality and size.
- Levalloisian flakes have a triangular or oval shape, but in Clactonian flakes, edges are not sharp. They require further retouch.
- In Levalloisian flakes, there is no loss of materials, but in Clactonian flakes there is loss of materials.
- In Levalloisian technique blow is given on prepared platform, but in Clactonian technique blow can be given anywhere on the margin of nodules.

Discoid Core or Mousterian Technique: Tools produced by the techniques have been reported from the French cave site of La Mousier in a small village on the Vezere river in Dordogne, France. It is a factory site where large number of cores and flakes were prepared by this technique. From the tools reported from this site, this technique has been reconstructed.

This technique involves the selection of a large flake with a flat surface on one side, or a flat nodule, or a core with one of its sides

Fig. 16.10: Discoid or Mousterian Core

being flat. Now, from the flat surface of nodule or flake, small flakes are removed. These flakes look round or polygonal leaving identical flakes scars on the core.

In the next stage, similar small flakes are removed from the other side of the core depending on its size. The residual core shows a central flat surface and bevelled rim. This is produced by the intersection of flake scars and cortexed surface of the nodule.

Levalloisian cores and Discoid cores and also flakes produced from them show differences and advantages. In case of Levalloisian core, only one flake can be produced at a time. For the production of another flake, the core has to be prepared again. In discoid core, several flakes can be removed without preparation of the core. The tortoise core can yield large flakes ranging from 5" to 7" in length at times. The discord cores are smaller in size ranging from 2" to 4" in length. Levalloisian flakes were converted into beautiful ovate hand axes which have been reported from different parts of Western Europe, West and Central Asia and North Africa. Discord cores were used for making tools like scrappers and points. Levalloisian flakes show a number of small flakes scars on reverse side indicating the previous preparation of the core. Discoid core shows only three or four scars or none.

Blade or Fluting Technique: This is a new technique of producing stone tools. The pre-historic man had begun to produce tools by this technique from middle Palaeolithic period. This technique has been regularly and extensively used during upper Palaeolithic, Mesolithic, Neolithic and Chalcolithic periods. By this technique, long, narrow, and parallel sided flakes were produced in different parts of the old world.

It is the fitness of the discussion to present here the difference between flake and blade. A blade is a narrow flake with nearly parallel sides and mostly thin and flat surface. By this definition, all flakes cannot become blades, but all blades are flakes.

Unlike different flake producing techniques discussed earlier, in blade producing technique the raw materials used and nature of core are different. The cores are generally cylindrical and fluted. The flakes are long as well as narrow and thin with faceted platforms. The nature of raw materials is dependent upon needs of blade produced. For blade production, materials were fine-grained such as jasper, chert, chalcedony, agate, obsidian belonging to silica group. In this technique blades were not produced directly from the cylindrical core. Rather blades were produced by employing a wood or bone punch placed over the core and exerting pressure in different ways. The technique of removing blades by direct percussion has also been reported from American Indians.

The careful observation of cores and blades reveal that pre-historic man had used pressure technique, indirect or direct percussion technique or only percussion technique for the purpose of manufacturing blades of different dimensions.

Pressure Technique: While employing this technique of producing blades, the pre-historic man had to follow a number of steps. The first step in this technique was preparation of a core. Here, tabular or cylindrical cores were roughened by other hard semiprecious stones or by an abrasive like sand. The roughening of core was done to facilitate the producer of the tool to hold the punch against the platform of the core and not to allow it to fall down.

Fig. 16.11: Production of Blades by Pressure Technique

In the second stage of preparation of core, small platforms were made along the edge by removing small chips so that it could become faceted. In certain cases, the core required further preparation particularly when crested ridge was made. For this purpose, a number of flakes were removed along the longitudinal axis alternatively and at right angles to its face to form a ridge with a zig-zag way. This kind of flakes create weaknesses parallel to the ridge and facilitate the removal of long flakes from the core. The crested ridge acts as a keel to hold the core in position at the time of flaking. Cores and flakes with such preparations have been found during Neolithic period in France and Chalcolithic period in India.

In the next stage a stick or a shaft measuring 2.5' to 4' in length and 2" to 3" in thickness was selected. Its lower end was fixed at a pointed bone or horn. The other end was provided with a cross piece against which producer's chest was rested. The maker of tool kept the core between his feet in standing or sitting position. Finally, the maker held the staff with both hands, rested his chest against the cross of staff and applied impulsive pressure. As a result of this pressure technique, a long flake measuring 10" to 12" was produced. These flakes were converted into spear had and arrow head.

Percussion Technique: In this technique of blade production, surface roughening is done on the core following by the preparation of a small platform at one end. Finally, a short wood punch is placed against it. A tap is given by a mallet for the removal of blade. This technique was tried on obsidian by two Frenchmen named A. Cabrol and L. Coutier. They had produced blade by this technique. In this technique as practised in Peru a bone wedge was placed across the platform and struck lightly to produce the blade. This technique was attempted by Tylor in Peru. Barnes had used this technique on glass. He had produced a narrow blade of 12" length. In Western Europe, percussion technique of producing blades has been reported since Acheulian times and continued till upper Palaeolithic period. This technique was followed successfully during Mesolithic, Neolithic and Chalcolithic period. Short blades were produced during these periods.

The cores used for blade production during these periods were small ranging from 0.5" to 2.5" in length and were of three types: I. Pointed or conical, II. Flat based and III Chisel or oblique ended.

The steps involved in this technique included the selection of a cylindrical or quadrilateral core, removal of smoothness from the surface of the core by roughening, preparation of platform or faceting by removal of small chips, making of crested ridge for facilitation flaking and for tight fixation on the ground or anvil and removal of blade flakes by pressure percussion with the help of a punch and light hammer. After removing flakes from one side of the core, another side was selected for blade production.

It appears that pre-historic man used pressure technique, pressure and indirect or direct percussion technique, or only percussion technique in production of blade.

Apart from the above technique, the technique of pressure flaking by a *fabricator* or by a *lame escaitee* particularly in case of baked blade by trimming along their backs and microliths by applying pressure. Very small flat flakes were also removed by this technique.

The technique of pressure flaking by finger nails was employed for producing small flakes on blades with finger nails. The example of this technique is production of saw with beautiful denticulations along the edge during Mesolithic and later cultural periods.

The pressure flaking by slit method was done for the production of microlithic burine by Mesolithic man.

Retouch or Secondary Working: This technique is applied on artefact for giving a sharp working edge or for creating a smooth surface after primary flaking. That is why this technique is known as retouch or secondary working technique. Pebble tools, hand axes and cleavers of Lower Palaeolithic were not subjected to secondary working. But tools of succeeding periods underwent secondary working after primary flaking or removal from the core. Secondary working was done on certain margins or sometimes all around the implement. Secondary working was also applied to the butt end of the tool to facilitate hafting. Hafting was also a kind of retouch technique. This retouch as a technique of tool production was employed to give sharpness to the tool or to strengthen the tool. Some raw materials like quartzite, dolerite whose original edges were sharp as deliberately made by man. But retouch or secondary working technique was discovered to achieve workable edge, when the tool used to become useless due to working.

Fig. 16.12: Pointed Core

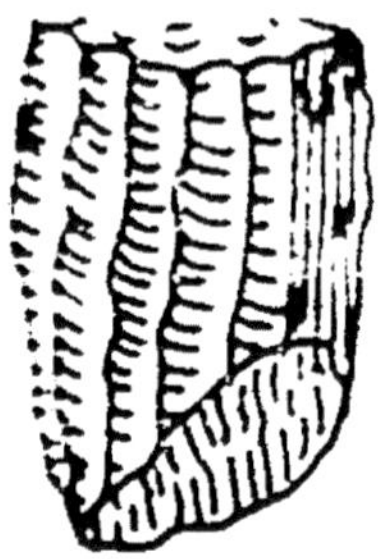

Fig. 16.13: Chisel or Oblique Ended Core

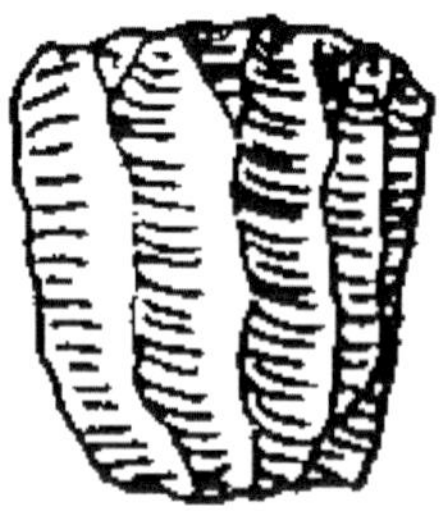

Fig. 16.14: Flat Based Core

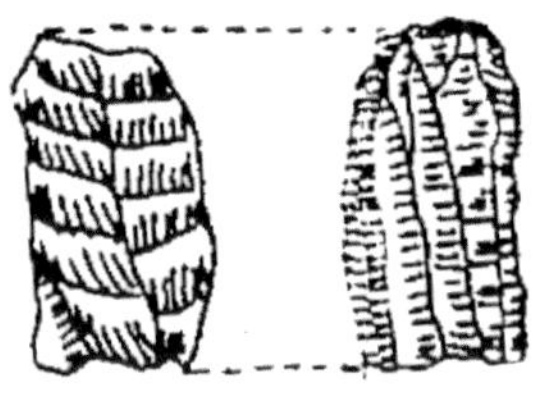

Fig. 16.15: Core with Cross Flaking and Crested Ridge

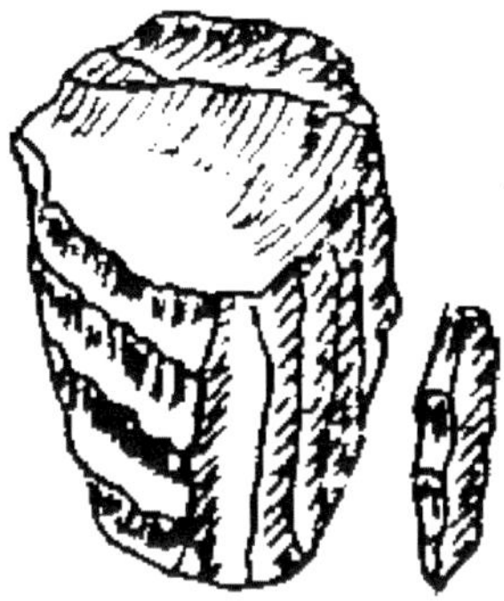

Fig. 16.16: Core with Parallel and Cross Flaking

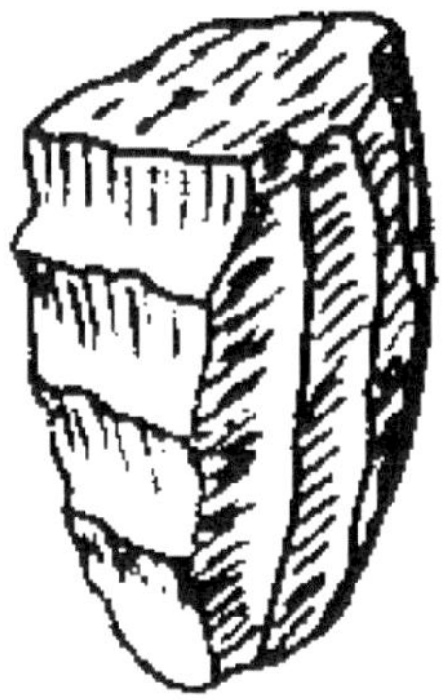

Fig. 16.17: Truncated Core

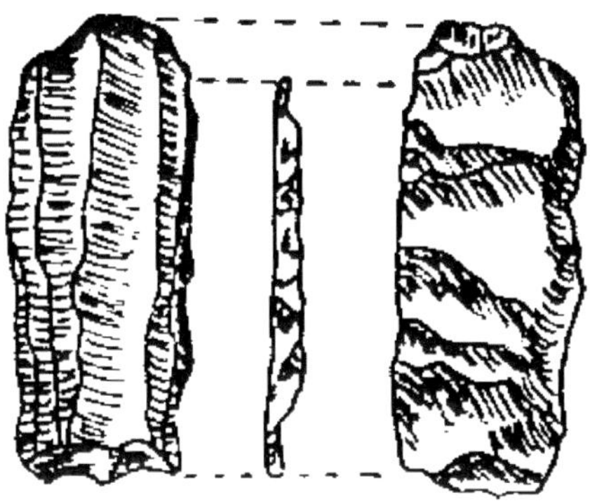

Fig. 16.18: Core with Ripple Flaking

Fig. 16.19: Nibbling Retouch

Fig. 16.20: Oblique Retouch

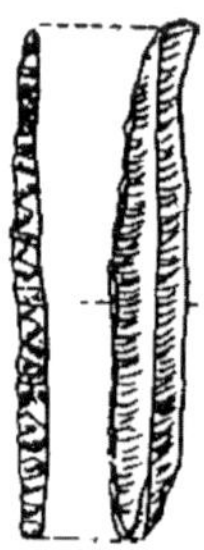

Fig. 16.21: Step Retouch

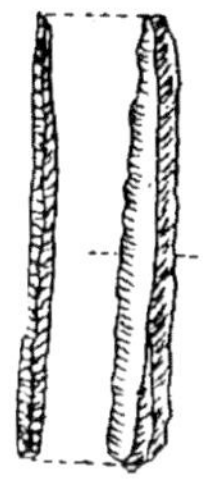

Fig. 16.22: Ridge Back Retouch

Types of Retouch Technique: On the basis of extent and nature, retouch technique may be named as following types:

- Nibbing Retouch. In this type, tool is retouched by removing small indentions here and there quite carelessly.
- Oblique Retouch. In this type, retouching on tool is carried out at an angle.
- Straight Retouch. In this type, retouching is done regularly and systematically.

Technique of Executing Retouch: Retouch was executed by the following techniques:

- Holding the edge against a stone or wood. In this technique retouching was done by holding the edge of tool against a fixed piece of stone or wood and moving it up and down carefully.
- Striking with a small hammer stone. In this technique, retouching was done by striking the edge of the tool with a hammer stone.
- With the help of a pointed instrument. In this technique, the edge of tool was retouched with the help of a sharp, thin and pointed instrument like tooth of walrus.

Retouch technique appeared for the first time on the Abbevillian-Acheulian type of hand axes and Mousterian points in Europe. This technique became well known in upper Palaeolithic period and continued during Mesolithic period in the production of different types of microliths. By the appearance of Chalcolithic and Bronze ages, the use of this technique began to decline due to massive production of parallel sided blades. In these periods, this technique was confined only to the production of lunates, trapezes and blunted back, knife blades and tanged specimen.

Grinding Technique: During Palaeolithic and Mesolithic periods tools were prepared by the technique of flaking or chipping. Surfaces of tools remained coarse and uneven. No attempt was made for making the surfaces of tool smooth and even. In the Neolithic age, the technique of grinding emerged in the history of mankind. On the basis of evidences of ground axe in association with Solutrean and other upper Palaeolithic cultural traditions of Western Europe and appearance of a rounded pebble converted into an axe by little grinding from Danish Kitchenmidens. Coughlan (1943) hints that this technique of grinding might have developed from a Palaeolithic sharp edged scrappers. The origin of this technique is also associated with nature of raw materials and physical environment. The rocks on which the technique of grinding was used were all fine grained igneous rocks such as bosalt, dolerite, epidolerite, etc. Metamorphic rocks were also used to a limited extent. This technique was never attempted on silicons rocks. The igneous rock was preferred because of pressure of joint planes on them which were helpful in converting the tool according to the plan of makers.

The Neolithic cultures of South, North and Eastern India bear testimony regarding the use of grinding technique in manufacturing tools and weapons.

Allchin (1960) has recognised five stages of grinding technique. These stages are as follows:

- Primary Flaking
- Secondary Flaking
- Pecking or Hammer Dressing
- Edge Grinding
- Overall Grinding or Polishing Technique

- *Primary Flaking*: This is the first stage of the grinding technique. It involves initial flaking of a pebble or nodule of any fine grained rocks in a desired shape of the tool. This is done by employing block on blocks technique or direct percussion. A Spherial or discoid hand hammer is used for this purpose.

- *Secondary Flaking*: This is the second stage of the grinding technique. This is done after primary flaking. After primary flaking, line flaking is done to regularise the form and make the edge sharp. This is called secondary flaking. After secondary flaking tool looks like an Abbevillian hand axe.
- *Pecking or Hammer Dressing*: This is third stage of grinding technique. This stage involves the pecking of tool by battering or hammering. This is done by removing projections, ridges and depressions produced by flaking. For this purpose a pointed instrument is used. In absence of pointed instrument, a small discoid or cylindrical instrument is used for this purpose. The tools with pecking present even surfaces.
- *Edge Grinding*: This is the fourth stage associated with grinding technique. In this stage, the tool is placed on a big bounder or stone and is rubbed by moving it up and down. Rubbing is done with the help of sand or other coarse materials. Some water was used as abrasive. The groves on the boulder and smooth surfaces on tool indicate that grinding was usually done on edge portion. That is why this technique is also known as edge grinding technique.
- *Overall Grinding or Polishing Technique*: This is the fifth and final stage of grinding stone tool technique. In this stage, grinding was done to achieve very fine smooth surface of the tool. This was done with the help of using some grease or oily substance. Some sands were mixed in grease or oily substance then grinding was done with intense rubbing.

Evolution of Tool Types

Evolution of tool types is also closely related to raw material available in the environment of pre-historic man and technique of their production. For the purpose of study, evolution of tool types can be discussed under the following headings:

- Tools of Lower Palaeolithic Culture
- Tools of Middle Palaeolithic Period
- Tools of Upper Palaeolithic Period

- Tools of Mesolithic Period
- Tools of Neolithic Period

Tools of Lower Palaeolithic Culture

The tools of lower Palaeolithic culture may be discussed under the following sub-headings:

Pebble Tools: The literally meaning of pebble tool is any tool made on a pebble. But in practice the term pebble tool is applied to a variety of choppers, scrappers and hand adzes whose working edge is prepared by block-on-anvil technique and remaining part of the tool retains original surface. These tools are quite big and massive. They are characteristics of lower Palaeolithic culture of South-East-Asia (Burma), North West India and East Africa. They were discovered for the first time at a site on river Kabul (Uganda), at Olduvai (Northern Tanganyika) and some sites on Soan and other rivers in Western Punjab. Such types of tools were also reported from the Vaal river in South Africa, at cave sites Makapansgat and Sterkfontein in Transvaal, on some sites in Algeria, East Punjab and a large number of sites in Peninsular India.

At the sites of Olduvai and Sterkfontein, pebble tools have been discovered from a separate stratigraphical layer. They have been considered as belonging to a separate cultural tradition known as pebble tool culture. They were associated with hand axes made on nodules and flakes. Deshmond Clark (1962) suggested the term Oldowan culture in place of pebble culture and tools of such culture as Oldowan tool in place of pebble tool. The antiquity of this tradition is estimated to be about two million years in Africa, whereas in Java and China it may be about half million years old. Pebble tools were found in association with flakes and tools made on other rocks, therefore, some scholars argue that they do not indicate a separate culture. But it is clear that pebble tools found in above areas are not of same antiquity. Recently pebble tools have been discovered from Hungary, South France. In India pebble tools form part of hand axe-cleaver industry. Various types of pebble tools are as follows:

Chopper: This is a large massive pebble core tool. It is prepared by unifacial flaking of the terminal end. In some rare cases the flaking might extend over one of the surfaces but do not include the butt end.

The butt end is kept untouched as a rule. A chopper can have U shape or even circular appearance. That is why working end may have a straight and transverse border to a semi-circular border. If two terminals of the working border are joined by a straight line, the maximum thickness of a chopper usually falls posterior to this line, i.e. towards the butt end. If the maximum thickness lies anterior to this line, it is advisable to consider such a specimen as a flake core. Choppers must have been used for scrapping and clearing purposes.

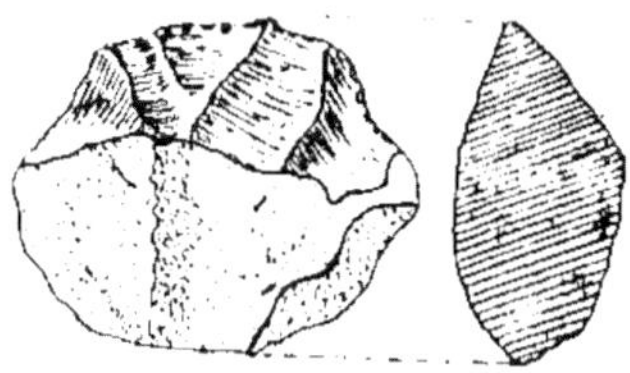

Fig. 16.23: Chopper Tool

Chopping: This is only a variation of the chopper. The only exception in this tool is that here flaking is done from both the surfaces. Terminal flakings are done alternately by altering the surface facing the worker. As a result of this the working end of the chopping tool is jagged or wavy in contrast to sharp border of a chopper. Here, also the shift of maximum thickness towards the working border can be taken to include them as flake core. Here, the remaining surface including thick butt end retains original surface or cortex.

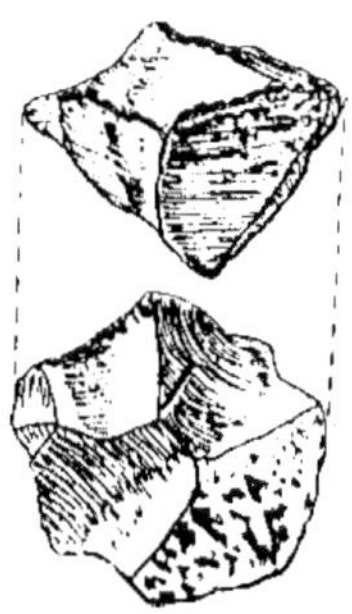

Fig. 16.24: Chopping Tool

The term chopper and chopping tools were first suggested by H.J. Movius in 1944 while he was analysing Palaeolithic tools from India.

These tools appeared for the first time in lower Palaeolithic period and later in middle Palaeolithic period when their size has become smaller.

Most of chopper and chopping tools possess a transverse cutting edge. But when the cutting edge tends to become pointed, they have been called as pointed oblates by Paterson and Drummond in 1962 when they were analysing Indian Palaeoliths. Sen has called choppers as unificial scrappers and chopping tools as bifacial scrappers. Seiveking has termed them as unifacial cleavers, and bifacial cleavers.

Both chopper and chopping tools are usually prepared by primary flakings alone. They are delivered by stone hammer or block-on-blocks technique. In some special cases secondary flaking was also done. But when there is continuous retouching along the border, then they are termed as core scrappers.

Scrapper: This is supposed to be a smaller pebble tool than chopper and chopping tools. They may be unifacial or bifacial. The unworked side serves the purpose of butt end.

Rostro-carinate: This is a pebble tool with cortex at the posterior end and partly on lateral sides. Here, flaking is done in such a way that a large flake comes out from the under surface, while upper surface bears partial flaking with a ridge or keel. The tool looks like a beak when viewed from front and top.

Proto Hand Axe: This pebble tool is a variant of rostrocarinate. Its one of surfaces is partially flaked to give a cutting edge. Proto hand axe and rastro-carinate are pebble butt hand axes. But proto hand axe is better shaped with more flaking. Secondary work is rare in both cases.

Clactonian Core and Flake: Clactonian core is a core or a primary piece of rock whose flakes were removed by Clactonian technique. The resultant flake shows irregular shape, large size, prominent bulb of percussion and semicone formed by the ripples of force.

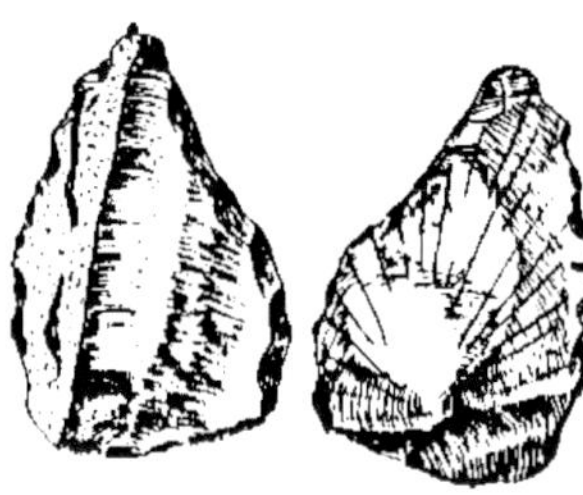

Fig. 16.25: Clactonian Core and Flake

Hand Axes: Hand axes are bifacially flaked core tools. The have biconvex profile. They possess thick and heavy butt end and a thin tapering pointed working end. The tapering end may be with or without unflaked portions. Hand axes are also known as bifaces or cou-de-pong. A French pre-historic anthropologist, Bucher de Perthes, was the first pre-historian to report about hand axes. After him, hand axes were reported from Europe, Africa and Asia particularly in India. Hand axes of those days do not resemble with axes of present day. Hand axes have one end pointed and other end thick and blunt. The pointed end is used for digging and sides are used for cutting or splitting. That is why hand axes are also known as multipurpose tools.

As hand axes were reported first of all from Olduvai George in East Africa, it is believed that they were first made from pebble. Later on they spread to Europe by diffusion. They differ from hand axes of other parts of the world. Hand axes have been placed under three traditions on the basis of method of manufacturing. These traditions are Chellean, Abbevillian and Acheulian. The crude variety of hand axes were put under Challean tradition. The finest variety of hand axes were placed under Acheulian tradition. In between these two traditions were placed hand axes of Abbevillian tradition. As there is not much difference between Chellean and Abbevillian hand axes, therefore, they can be placed together.

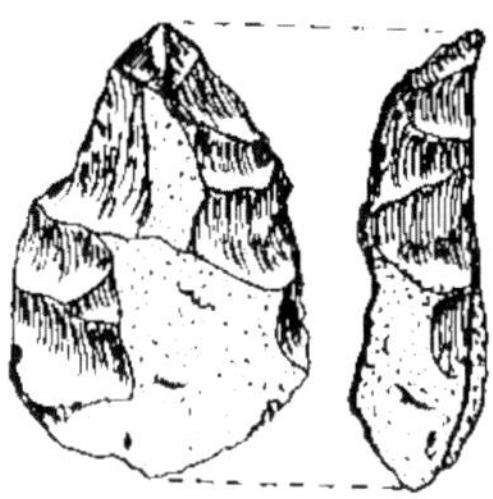

Fig. 16.26: Proto Hand Axe

Chellian-Abbevillian Hand Axes: These hand axes show a deep and wide flake scars at one end or side or in semi-circular area of tool. This kind of flaking was done by block-on-black technique or stone hammer technique. The outlinen of these hand axes are irregular. The working end is zig-zag. The cortex bears thick butt end. These tools are crude in nature. They have been flaked from upper and lower surfaces which intersect to form a working edge for use. The name Chellian has been given after a type site chelles on the junction of river Siini and Marne. The name Abbevillian was given after a type site Abbevilles on the Some river. Both these type-sites are situated in France.

Fig. 16.28: Abbevillian Hand Axe

Acheullian Hand Axes: The name Acheullian has been given after a type-site named St. Acheul in the Some valley of France. This type-site was discovered by Boucher de Perthes in 1836. Hand axes discovered from this site are known as Acheullian hand axes. These hand axes are regular in outline and beautiful to look at. They were produced by removing thin flakes on both surfaces. In cross section they are biconvex or lenticulate. They would have been produced by a light cylindrical hammer of wood, bone or stone. They show technological development in hand axe industries. Acheulian hand axes can be divided into following sub-types:

- *Peer-shaped or Pyriform Hand Axes*. As these types of hand axes resemble with pear in shape, that is why they are called as pear-shaped hand axes or pyriform hand axes. These stone tools are short and heavy. Their working points are rounded. The general technique employed to receive the shape of these tool is primary flaking followed by few secondary flaking to sharpen the edge.

Fig. 16.27: Pear Shaped Hand Axe

- *Ovates*: These stone tools are oval in outline. That is why they are known as ovates. These are thin and symmetrical tools with biconvex profile. Lateral sides and working end of these tools continue to form a ridge across the butt end. All over the border of the tool secondary working is done to achieve a working edge.

Fig. 16.31: Ovate Hand Axe

- *Cordiform or heart shaped hand axes*: As these hand axes resembles with a heart in shape, they are called as cordiform or heart-shaped hand axes. These tools are symmetrical with a broad round butt end. Gently incurved sides converge into a point. They appear biconvex or lenticular in shape.

Fig. 16.32: Cordate Hand Axe

- Lanceolate hand axes: These tools have a fairly long tapering sides which end into a pointed end. The surfaces of these tools are flat. The width of the tool is less than length as in case of a lane-head. These are well-planned, well-finished and symmetrical tools.

Fig. 16.30: Lanceolate Hand Axe

- *Triangular hand axes*: The shape of these tools are triangular. Their sides meet like a triangle. Their surfaces are flat. In section they appear as triangular (Fig. 16.29).

Fig. 16.29: Triangular Hand Axe

- *Micoquian hand axes*: These are small tools. Their shape is triangular. The working end of these tools is elongated. Like Acheulian hand axes they are not so symmetrical. The butt end of the tool is thick and heavy. The butt end preserves the original surface. Unlike Abbevillian hand axes, they are finely retouched. These tools were reported for the first time from a French site named La Micoque. That is why they are known as Micoquian hand axes. Stratigraphically this follows Acheullian types.

Fig. 16.33: Micoquin Hand Axe

- *Unifacial hand axes*: The Chellian-Abbevillian and Acheullian hand axes described above are generally known as bifaces or bifacial hand axes. But in India, some of hand axes of Acheullian industry reveal flake scars only on one surface. They are called as Unifacial hand axes. They have been produced from large flakes not on pebbles or nodules as in case of bifaces. In case of Unifacial hand axes, one finds that only upper surface is worked by removing small shallow flakes. Here the under surface contains a plain flake surface. The under surface intersects with the upper surface to form a thin convenient cutting edge.

As hand axes were first developed stone tools, they represent a universal and unspecialised tool. They are multipurpose tool. They were used for cutting, digging, splitting, scrapping and chopping. They were in use for longer time period in the history of mankind.

In England, a new industry called Clactonian industry has been recognised. This industry is quite different from pebble tool hand axe industry of upper Palaeolithic period. Flakes produced by Clactonian technique reveal that they are not by products of hand axe industry. They appear to have been removed from big stone blocks. They possess a wide striking platform and prominent bulbs of percussion with an angle of about 120° between the two. Their upper surfaces retain cortex, while the sides have been secondarily worked to convert them into points and scrappers. Such flakes occur in association with hand axes or without hand axes as in England.

Cleaver: It is a cutting tool. It is also known as cleaving tool. Hence, it is called as cleaver. It is prototype of an iron axe of present day. The cutting edge of the tool is broad. It is found in association with hand axes, but its antiquity is later than the hand axes. For the first time cleavers were found in association with middle Acheullian hand axes. They were recovered from the bed III of Olduvai George in East Africa. This is a flat tool. It is made on a broad rectangular or rarely triangular, or convex side or end flake. Sometimes cleavers are produced from core or pebble butt or from a thick flake with similar feature. Generally, the under surface of the cleaver retains the primary flake scar and upper surface retains a portion of original surface. The upper surface is chipped all over. The final large flat scar removed at the broad end of the flake intersects the scar of under surface to give rise to a broad and straight cutting edge. The two thick longitudinal sides are trimmed to facilitate holding or hafting. Retouching is done on bifacially worked thick flakes extensive to give a U or V-shape to the entire tool. The broad straight edge in front is thinned out by removing a single flake by giving a controlled horizontal blow. In such instances the resulted scars are known as cleaver facets. The technique of removing single large flake from the under surface in the preparation of hand axes and cleavers involves the preparation of core and platform as in Levalloisian technique. It was first noticed at the Victoric West in South Africa. That is why it is named after the site. Such type of cleavers have also been reported from India.

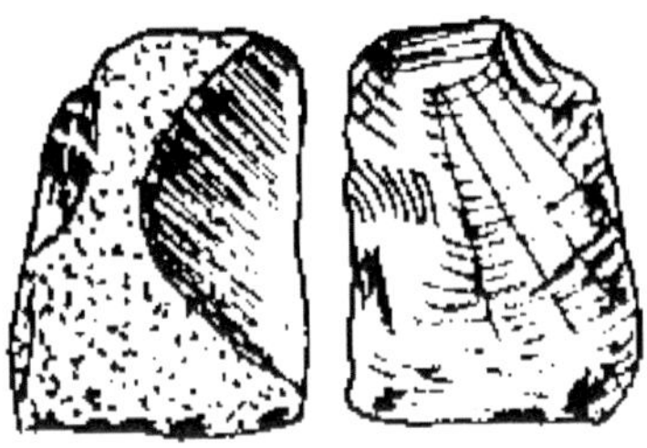

Fig. 16.34: Cleaver

Cleavers may be divided into following sub-types on the basis of shape or form of the butt or form of the edge or nature of cross section:

- Cleaver with square or rounded U-shaped straight broad edge. These are generally square or rectangular in shape.
- Cleaver with pointed butt and straight broad edges. These are generally V-shaped or roughly triangular in shape.
- Cleaver with broad and narrow butt.
- Cleavers with parallelogrammatic section.

Tools of Middle Palaeolithic Period

The lower Palaeolithic period was succeeded by middle Palaeolithic period. This period witnessed a change in the raw materials used for making tools, typology, technology, sizes of tool and environment. Tools were small in size and made on flakes of Levalloisian and Clactonian type of fine-grained metamorphic rocks of Silicious type. The technique of retouch is quite common on tools produced in this period. The main tool types of this period are:

- Scrapers
- Points
- Borers-Cum-Scrapers
- Small Choppers
- Burin

Scrapers

Scrapers are smaller tools made on flakes of medium size. These are thin tools. They are manipulated with fingers just as hand axes and cleavers are held in the palm and used. The tool scrapper is subjected to secondary working after primary flaking is done on the upper surface. The under surface bears a single flake scar of detachment. It is rarely retouched. Scrapers are used mainly for scraping bark of trees, dressing of wooden and bamboo shafts and skin of animals. Scrapers can be classified in a number of sub-types based on shape, position and nature or the edge for scraping. They are as follows:

- *Side scraper*: This is a sub-type of scraper in which one or both of longer sides from the upper or underside or sometimes

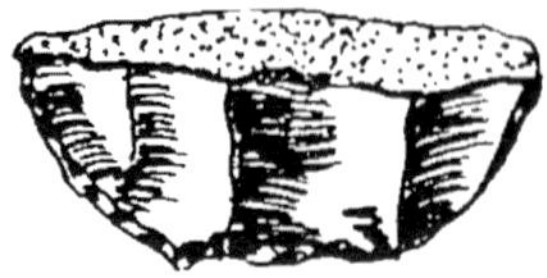

Fig. 16.35: Convex Scraper

Fig. 16.36: Concave Scraper

Fig. 16.37: Borer-cum-Scraper

from both surfaces retouching is done. That is why they are called as single side or double side scrapers. Working edges in both cases are straight more or less.

- *End scrapper*: In this sub-type of scraper, scraping edge is confined to the shorter side which is steeply retouched. As the edge is steep like, so it is known as end scraper.
- *Round scrapper*: In this sub-type of scraper, working is confined to all round the flake.
- *Concave or hollow scraper*: In this sub-type of scraper, working edge is intentionally made concave by flaking or by taking advantage of natural concavity. The tool is

retouched from the upper or under surfaces. Such tools began to appear from lower Palaeolithic period of Europe and India. But they become characteristic tool of middle Palaeolithic period. In some cases of these scrapers, concavities may be at more than one side resulting in the formations of pointed sides or ends in which case the nod is called ascraper-cum-borer.

- *Convex scarper*: In this sub-type of scraper, the working edge is convex or arched and obliquely retouched from above or under surface.
- *Concavo-convex scraper*: In this sub-type of scraper, the working edges are concavo-convex i.e. concave and convex respectively.
- *Side-cum-end scraper*: In this sub-type of scraper, one or two are longer sides and ends bear retouch. That is why it is known as side-cum-end scraper.

Points

These are made either on Levalloisian flakes or other simple flakes. The tools may be large or small, thick or thin and triangular to leaf shaped. Both longer sides of the tool bear retouch marks depending upon the extent of working. Points may be called as Unifacial or Bifacial. Functionally, thin, small, triangular or leaf shaped point may have been used as arrow heads. The large and thinner ones with mid ridge may have served the Javelin or even as spearheads. Points with incipient tang must have been hafted and used.

Fig. 16.38: Unifacial Point

Fig. 16.39: Tanged Point

Fig. 16.40: Bifacial Point

Borer or Awl

The two wings of this tool are intentionally retouched by deep notches. The body of the tool may be square, rectangular or even round on a flake or nodule where minimum working has been executed for converting them into borers or points.

Fig. 16.41: Borer or Awl

Small Choppers and Chopping

These tools bear unifacial and bifacial flaking respectively as their counterparts of pebble tool culture. But they are smaller in size.

Burin

Burins are also known as gravers. Burins are characteristic tool type of upper Palaeolithic period. They were mainly used for engraving on soft stone or bone as well as on walls of rock shelter and caves. Perhaps they were also used in making slots in wood and bone. These are special kind of tool fashioned by pre-historic man.

These are small chisel like tools produced on blade like flakes with a sharp but thick cutting edge formed by the intersection of the levelled or sloping surfaces. Burins can be classified into different types on the basis of extent and nature of working on tool. The most common type of burin is called Burin Bec-de-flute. This type of burins are made on blade flakes. The blade flake selected for the preparation of burin is trimmed a little on both sides at one end. As a result a rough point is produced. Now this rough point is again flaked to get a fine point at the tip of the flake blade. In this way burin Bec-de-flute is produced.

Fig. 16.42: Burin

Tools of Upper Palaeolithic Period

The upper Palaeolithic period is characterised by blade and burin industries of middle and late wurm glacial period of Pleistocene epoch. The makers of these industries were fully modern man. Besides blade and burin industries, the man of this period used to make tools of bone, horns, ivory and antlers. They lived in caves and used to do carvings and painting on rock shelters and caves. They used to do carvings also on bones. They used to bury their deads. They wore animal skin as hide. Their cultures have been reported from Europe, Africa, West Asia and India. The typical tool types of upper Palaeolithic period are as follows:

- Bifaces
- Burins
- Backed knives or blades with oblique blunted or strait blunted backs
- Knife points
- Scrapers
- Arrow-heads
- Piercers
- Lunates
- Trapezes
- Triangles

Tools of Mesolithic Period

Mesolithic period is a transitional stage between Mesolithic period and Neolithic period. This period made its beginning after the disappearance of the Pleistocene epoch. The man of this period produced compound tools called Microliths. Microliths were hafted in convenient handles and used for various purposes. Besides, the production of microlithic tools, the technique of blade production also continued in this period. Based on the morphology of stone tools, two categories have been deciphered. They are as follows:

- Non-geometric tools
- Geometric tools

Non-geometric tools

Non-geometric tool types are older than geometric tools. The different tool types of Mesolithic period are as follows:

- Cores
- Blades
- Microliths

Cores

Cores are of two types— (i) Flake core, and (ii) Blade core. Flake cores are ordinary cores which were not prepared for the production of flakes of any particular size and shape. They bear liberal flaking. The flaking is done from various points of the core. No prepared platform is present in such cores. Flake core occurs in small number. Blade cores are regular cores. They possess prepared platforms of which thin, narrow, parallel sided blades were removed. In shape, they are cylindrical, tabular, quadrilateral or sometimes even pyramidal or conical having pointed chisel or flat based end. Blades from such cores are removed from more than one platform and a maximum of three platform. Majority of cores measure between 2" and 4" in length.

Core rejuvenation flakes: These are flakes which have been removed from cores to improve the existing platform or to provide fresh striking platform when the possibility of detaching blades from one platform gets exhausted. These are removed from the apex, base and toe of cores. They retain terraces of blade scars and platform on their surfaces.

Chips: These are by-product or waste flakes resulted from the preparation of waste cores and various types of artefacts. They are small and irregular in shape. They do not bear any marks of use and retouch.

Flakes: These includes utilised and unutilised ones. They may be end, side or interminant flakes. They are long, oval and leaf like in form and sometimes irregular.

Blades

Blades are thin, long and parallel sided flakes. They may be retouched or unretouched. Majority of them belong to latter type. Only a small proportion include the former type. The retouched blades include those intentionally worked by secondary working for use. The retouching may be on one lateral side or both sides. The blades in general have one to two mid-ridges on the dorsal surfaces. The unretouched blades, on the other hand include utilised and unutilised types. Majority of them belong to utilised types. These blades measure from 2" to 4" depending on the length of the core.

Fig. 16.43: Single and Double Straight Sided and Straight and One End Pointed Blade

Fig. 16.44: Single and Double Straight Sided and Straight and One End Pointed Blade

Fig. 16.45: Single and Double Straight Sided and Straight and One End Pointed Blade

Fig. 16.46: Obliquely Blunted Pen Knife Blade

Microliths

These are usually finished tools. They are geometric and non-geometric type.

Geometric Tools

They include the following types:

Lunate or Crescents: These are small microliths. They are made on parallel sided blades. They resemble with half moon in shape. They have round back called arc and straight opposite side called Chord. The arc is thick and intentionally blunted by steep retouch to facilitate hafting in a handle. The chord is almost always unretouched. Therefore, arc retouched lunates are common. A true lunate is supposed to be symmetrical in shape. Some scholars make a distinction between a lunate and a crescent.

Fig. 16.47: Lunate

- *Triangles*: These are made on broken blades. The sharp edge forms the base. Longitudinal sides are blunted. They may be geometrical and non-geometrical.

Fig. 16.48: Regular and Irregular Triangles

Fig. 16.49: Regular and Irregular Triangles

Fig. 16.50: Regular and Irregular Triangles

Fig. 16.51: Regular and Irregular Triangles

- *Trapezes*: They resemble with geometric trapeze in which shorter three sides are retouched steeply. The biggest size remains as a sharp cutting edge. It may be taken as transitional form of lunate. In a true trapeze, two parallel sides of original blade remain untouched. Non-parallel sides are retouched.

Fig. 16.52: Different Types of Trapezes

Fig. 16.53: Different Types of Trapezes

Fig. 16.54: Different Types of Trapezes

Fig. 16.55: Different Types of Trapezes

Fig. 16.56: Trapezoids

Fig. 16.57: Trapezoids

Fig. 16.58: Trapezoids

- *Transverse arrow heads*: In this type, the length between the cutting edge and its posterior border is more than between the lateral sides.

Fig. 16.59: Transverse Arrow head

Fig. 16.60: Obliquely Blunted Point

Non-geometric Tools

Non-geometric tools are of following types:

- *Backed blades*: These are parallel sided blades. They have one or both lateral sides retouched for the purpose of cutting.

Fig. 16.61: Truncated Blade

Fig. 16.62: Truncated Blade

Fig. 16.63: Truncated Blade

- *Obliquely blunted blades*: These are also known as pen knife blades. They possess a steeply blunted side which curves to meet the thin unretouched edge which acts as working edge. Blunting may be done on left or right side. Blunting may be partial or complete. The working edge is concave or concavo-convex and straight.
- *Truncated blades*: These are blades the broken ends of which are trimmed either transversely or obliquely to produce a scraping edge. The truncation is done at one or both ends. The working edge is transverse or straight.
- *Tranchets*: These are flake tools. Here, cutting edge is formed by the interaction of two or more flake scars from the two surfaces of the tool.
- *Hollow-based points*: In this type of tool a part of one side is at least steeply blunted. The base is intentionally hollowed by retouch. They include symmetrical and asymmetrical both types.
- *Micro-burins*: They are usually prepared on small cylindrical nodules. Around the central region a notch is prepared in the form of a wedge to a length of one-half of the nodule. The remaining half is broken by a single stroke. The smaller broken piece becomes the burin. As these burins differ in shape, size and method of manufacturing from Palaeolithic period, they are called as Micro-burins.
- *Scrapers*: Scrapers are of various sub-types such as side, convex, concave and round scrapers. They have been described earlier.

Fig. 16.64: Side Scraper

Fig. 16.65: End Scraper

Fig. 16.66: Sleep Ended Scraper

Fig. 16.67: Hollow or Concave Scrapper

Fig. 16.68: Convex Scraper

Tools of Neolithic Period

In Neolithic period one finds the existence of two kinds of industries:

(i) Pecked and ground stone industry and (ii) Blade and microlithic industry.

Pecked and ground industry: This Neolithic industry is characterised by the existence and production of following tool types:

- Celt or axes
- Adzes
- Chisels
- Wedges
- Scrapers
- Picks
- Borers
- Grinding or rubbing stones
- Saddle querns or mill stones
- Fabricators or hammer stones
- Sling stones
- Mace head or ring stones
- Axe hammers
- Flakes—simple flakes
- Flake blade

Celts or Axes: These Neolithic tools are mostly triangular in shape. They have broad sharp cutting edge with a butt. The butt may be pointed, round, blunt, broad thick or broad thin. They constitute the most significant and major group of the pecked and ground stone industry. They vary in size from small to large specimens. But majority of them are medium sized. Small and large axe are quite rare.

Fig. 16.69: Ground Axe in I Stage of Preparation

Fig. 16.70: Ground Axe in II Stage of Preparation

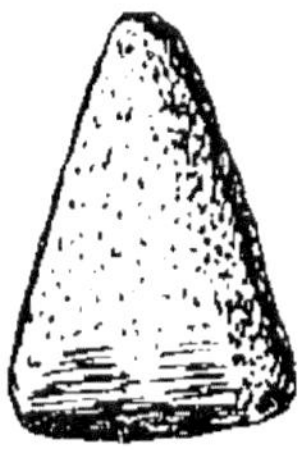

Fig. 16.71: Ground Axe in III Stage of Preparation

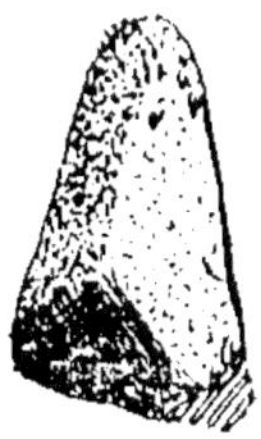

Fig. 16.72: Fully Ground Axe in Stage IV

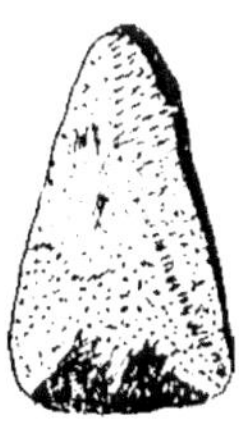

Fig. 16.73: Pointed Butt Fully Ground Axe

The materials used for making celts would have been procured from locally found dykes and sills in the form of fragments which vary from very fine grained basalts to coarse grained dolerites. At times diorite, granite, granodinorite and schists were also used. Evidences in the form of rock grooves or granite boulders were reported from the vicinity of water ponds and springs. These evidences reveal that axes were ground at their edges. These evidences show that Neolithic man had used pond water and some sand like abrasive for grinding and polishing the axes. Even today the village folk use the same technique for sharpening the iron made axes and blades. Working edge or blade of axes is often curved with slight or prominent convexity. Axes with oblique to straight blades also occur at times. Their bodies are thick with round bevelled and square sides. Majority of axes are oval in cross section at the butt end in the middle, while at the cutting edge lenticular section is most common. In longitudinal or vertical section axes are often flat oval or lenticular through plano-convex and concave-convex sections are also encountered. There is also direct evidence as to how axes were used by the makers. They were held either in naked hands or hafted into some kind of handle. Axes were used for clearing the forest for agricultural operation. Completely ground and polished axes are few in number. They were precious possession of Neolithic people. They were used mostly for ceremonial purposes.

There are also a special type of tools of Neolithic period. They are called as 'shouldered tool'. They have been reported from South East Asia, Burma and Eastern India. This tool has taken at its butt end for hafting purposes. The other end is cutting edge. Their body are square shaped.

Fig. 16.80: Shouldered Tool

Adzes: It is a tool for chipping or slicing away the surface of the wood. The cutting edge stands transversely, i.e. at right angle to the handle. Its bevel is ground on the inner face only. The entire outer face is slightly rounded. These are thin tools. They are triangular in shape. They are made on flakes. They have one flat surface and other slightly convex surface with a bevelled central edge. Adzes are hafted in such a way that the blade is at right angles to the handle. Usually bevelled edge is ground. It acts as a god tool of carpenter. It is believed to have been used in agricultural operation. Adzes blades are rare in peninsular India. They are common in Assam Neolithic sites of India, China and South East Asia. They are also called as Faceted tool of multipurpose nature. The first hafted specimen was reported from Iran at Arpachiyab.

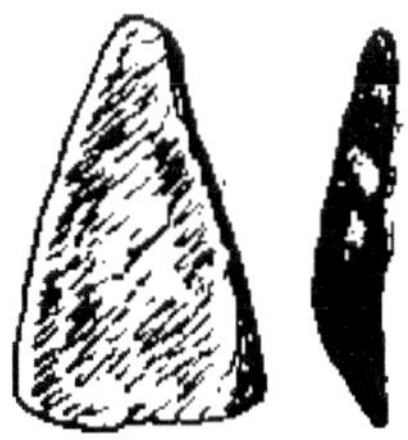

Fig. 16.74: Adze

Chisels. Chisels are narrow elongated cylindrical or rectangular axes with ground edges which may be straight or convex. They are the actual prototype of metal made Chisels used in carpentry. Their working edge is truncated biconvex which are more akin to those wedges. They might have been used for splitting purpose like wedges. They were also used for cutting purposes. Chisels like adzes are of rare occurrence.

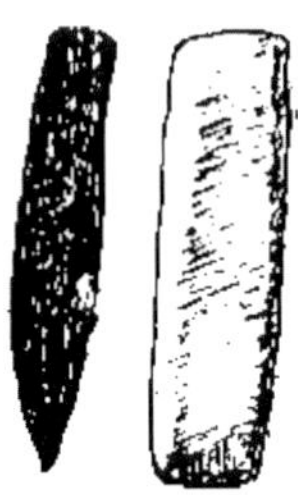

Fig. 16.75: Chisels with Square and Circular Sections

Fig. 16.76: Chisels with Square and Circular Sections

Wedges. These are small tools. They are roughly triangular, quadrilateral pieces. They have ground edge and packed surface. They were used for splitting wood. Probably, they were made from broken axes.

Scrapers: These tools were used for scraping skin and bark of trees. They are made on cores as well as flake. They can be divided into side, end, hollow, convex, concavo-convex, round or discoid, etc. They are of various sizes. They were found in association with Neolithic pecked and ground stone industry.

Pecks: They are irregular and pointed tool. They are prepared on cores by flaking. Some of them show pecking and grinding, too. Pecking has been done probably to facilitate hafting of the butt into a wooden handle. The grinding has been done to make the point sharper. These tools were used in digging and loosening soil and in agricultural activities. These tools have also been found occasionally on Neolithic sites.

Borers: Like pecks, borers are a variety of pointed tools made on flakes. They, too, occur occasionally at Neolithic sites.

Grinding or Rubbing Stones: These are a class of domestic implements. They are usually found in association with querns. They serve the purpose of grinding and pounding grains. They are made on small natural slabs in relation to the querns made on huge boulders. They may be oblong, rectangular, oval or circular in shape. Usually they do not show any sign of flaking but they are pecked all over with stone chisels to give them a flat surface and to produce dentitions for grinding. Grinding was not involved in their manufacturing. Robert Bruce Foote has called them as mealing stones. They have been found commonly on all Neolithic sites. Granite of various shades was the chief raw material used in the making of these tools. In vertical and horizontal cross sections, they are plano-convex, oblong, rectangular, oval, etc. There are various types of rubbing stones. The chief among them are those which have been worked and used on one surface only with signs of utilisation at one or both ends and on one on both sides.

Fig. 16.79: Muller

Saddle querns or mill stones: These tools have derived their names from their appearance to the ridding saddles. They were used for grinding and pounding grain and other cereals. They are rectangular, square and less frequently round in shape. They are made of large granite boulders. Foote has called them as 'mealing troughs'. They have been reported from all Neolithic sites. Majority of them are long, broad and shallow in depth. Their surfaces were made hollowed by pecking. This became deeper due to use.

Fig. 16.78: Saddle Quern or Mill Stone

Fabricator or Hammer Stones: These tools are round or cylindrical in shape. They were put to use in manufacture of blade tools and in making grinding or rubbing stones. Some of them were used on punches for dressing axes and similar ground axes. There are some hammer stone with a groove in the middle. They are called as Grooved hammer stones. They have been reported from Eastern India. The groove was used for hafting purposes.

Sling Stones: These are perfectly spherical in shape. They are made by pecking. Absence of battering marks on them suggests that they were the final product of the processes involved in the making of hammer stones. Some authorities think that they were used as weight. But it is more likely that they were used as sling stones for hunting depending on their size. They occur on all Neolithic sites.

Mace Heads or Ring Stones: These are thick massive circular stones with a well-drilled central hole. Their surfaces are sometimes pecked and ground. The central hole was pecked or drilled alternately from both surfaces and is of 2 to 3 cm in diameter. The diameter of hole narrows down from the surface to the centre. Their use as weights for digging sticks suggest that they were primitive agricultural implements. Their association with agricultural operations has been corroborated by Oakley in 1956. He reported that the Bushmen of South Africa chiefly utilised the perforated stones as weights for digging sticks.

Fig. 16.77: Ring Stone or Mace-head

Axe Hammers: Originally these were axes, but their cutting edges ceased to serve the purpose of axe due to prolonged utilisation. Therefore, they were utilised as hammers. They differ from hammers in having the shape and technology of axes though they are identical with hammers functionally. Due to their function they could be classified as hammer.

Flakes: These includes large flakes, small chips and irregular flakes which resulted during the manufacturing of finished tools. They may be end, side or interminate flakes. They may be rectangular discoidal triangular, oval and irregular in shape. They could be identified as simple flakes and flake blades.

Besides these tools, a number of miscellaneous tools with a multiplicity of functions occur with these tools.

Blade and Microlithic Industry

The blade and microlithic industry include the following types:

- Flake cores
- Blade cores
- Core rejuvenation flakes
- Plunging flakes
- Chips
- Utilised and unutilised flakes
- Unretouched blades

- Microlithc: Tools—Backed blades, obliquely blunted blades, truncated blades
- Microliths: Lunates, Trapezes
- Points
- Borers
- Scrapers of Various Types

Fig. 16.81: Sickle of Microlithic Blades Mounted into a Wooden Shaft

Fig. 16.82: Harpoon Formed Mounted Triangles

Fig. 16.83: Pointed Butt axe Hafted into a Stolled Handle

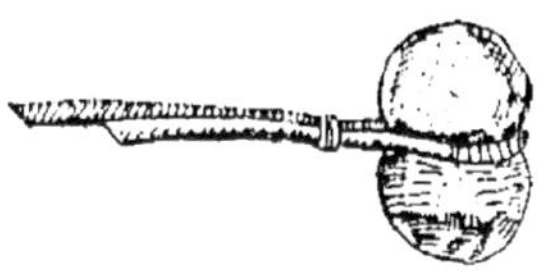

Fig. 16.84: Grooved Hammer Stone Hafted with a Bent Withy

Evolution of Axe

Fig. 16.85: Hand Axe of Lower Palaeolithic Period

Fig. 16.86: Polished Stone Axe Hafted in Wood of Neolithic Period

Fig. 16.87: Hollow Axe with Wooden Handle of Bronze Age

Fig. 16.88: Iron Axe of Romean Age Still Being Used in India

Bibliography

Beals, Ralph L. and Harry Hoijer (1965): *An Introduction to Anthropology*, 3rd ed. New York: The Free Press.

Bhattacharya, D.K. (1972): *Prehistoric Archaeologies*. New Delhi: Hindustan Publishing Corporation.

———, (1989): *An Outline of Indian Prehistory*. Delhi: Paloka Prakashan.

Binford, S.R. (1968): 'Early Upper Pleistocene Adaptations in the Levant'. *American Anthropologist* 70; 707-717.

Binford, S.R. and L.R. Binford (Eds.) (1968): *New Perspectives in Archaeology*, Aldine Publishing Company.

Birdsell J.B. (1972): *Human Evolution—An Introduction to New Physical Anthropology*, Chicago: Rand McNally & Company.

Bordes, Francis (1968): *The Old Stone Age*, J.E. Anderson Trans., New York: McGraw Hill Book Company.

Boyd, W.C. and M.F. Montague (1958): *Genetics and Races of Man*, Boston: Boston University Press.

Brace C.L., H. Nelson, N. Korn and M.L. Brace (1979): *Atlas of Human Evolution* (2nd Ed.), New York: Rineheart and Winston.

Brose, D.S. and M.H. Wolpoff (1971): 'Early Upper Palaeolithic Man, and Late Middle Palaeolithic Tools, *American Anthropologists*, 73: 1156-1194.

Butzer, K.W. (1964): *Environment and Archaeology*, London: Methuen and Company.

Campbell Bernard G. (1974): *Human Evolution: An Introduction to Man's Adaptation* (2nd Ed.) Chicago: Aldine Publishing Company.

Carter, C.O. (1970): *Human Heredity*. Penguin Books.

Ceghlan, H.H. (1943): 'Evolution of Axe', *Journal of Royal Anthropological Institute*, 73.

Chard, C.S. (1969): *Man in Pre-history*, New York: McGraw Hill & Book Company.

Charles Worth, J.K. (1966): *The Quaternary Era*, Vol. II, London: Edward Arnold Publishers Limited.

Child, V.G. [1951(a)]: *Social Evolution*, New York: Henry Schuman.

———, [1951(b)]: *Man Makes Himself*, London

———, [1951(b)]: *Piecing Together the Past*, New York: New American Library Mentor, Routledge and Kegan Paul.

Clark, Grahame (1960): *Archaeology and Society*, London: Methuen, University Paperback.

Clark, J.G.D. (1932): *Mesolithic Age in Britain*, Cambridge.

Clark, W.E. Le Gross (1949): *History of Primates*, London: Natural History Museum.

———, (1972): *The Fossil Evidence of Human Evolution*, Chicago, University of Chicago Press.

Coles, J.M. and E.S. Higgs (1969): *The Archaeology of Early Man*, London: Faber and Faber.

Daniel, Glyn (1950): *Hundred Years of Archaeology*, London

———, (1962): *The Idea of Pre-history*, London: CA, Watts and Company Ltd.

———, (1967): *The Origin and Growth of Archaeology*, London.

Darwin, C. (1859): *The Origin of Species*, New York: New American Library, Mentor.

———, (1871): *The Descent of Man and Selection Relation to Man*, New York, D. Appleti.

Das, B.M. (1980): *Outlines of Physical Anthropology*, Allahabad: Kitab Mahal.

——— (1981): *Micro Evolution Concept*, New Delhi: Concept Publishing Company.

David, P. (1972): *The Ascent of Man: An Introduction to Human Evolution*, New York: McMillan Company.

Days, M.H. (1988): *Guide to Fossil Man*, Chicago: University of Chicago Press.

Dixton, R.B. (1923): *The Racial History of Man*, New York.

Dobzhausky, Th. (1937): *Genetics and the Origin of Species*, New York: Columbia University of Press.

———, (1964): *Genetics and the Origin of Species*, Kharagpur, Oxford University Press.

———, (1970): *Genetics and Evolutionary Process*, New York: Columbia University Press.

Dunn, L.C. and Th. Dobzhausky (1952): *Heredity, A Race and Society*, New York: New American Library.

Eldredged, N. and L. Tattersal 1982: *The Myths of Human Evolution*, New York: Columbia University Press.

Else, J.G. and P.C. Lee (1986): *Primate Evolution*, Cambridge: Cambridge University Press.

Euner, F.E. (1958): *Dating the Past*, London: Methuen.

Flint, R.F. (1957): *Glacial and Pleistocene Geology*, New York: Wiley.

Garn, S.M. (1963): 'Culture and the Direction of Human Evolution', *Human Biology*: 35:221-231.

Gates, R. (1948): *Human Ancestry from Genetical Point of View*, Cambridge: Harvard University Press.

Goyal, Sri Ram (1961): *Pragaitihasik Manav Aur Sanskritiyan*. Gorakhpur: University Press.

Guha, B.S. (1935): *The Racial Elements in Indian Population*, Oxford, Oxford University Press.

Hadden, A.C. (1946): *Race of Man*: Cambridge University Press.

Haldane, J.B.S. (1932): *The Causes of Evolution*, London: Longmans.

Hays, H.R. (1958): *From Ape to Angel*: New York: Afreda, Knopf.

Hill, W.C.O. (1954): *Man's Ancestry: A Primer of Human Phylogeny*, London: Willmer, Brothers and Co. Pvt., Birkenheed.

Hole, Frank and R.F. Heizer (1965): *An Introduction to Pre-historic, Archaeology*, New York: Holt, Rinehart and Winston.

Hootan, E.A. (1954): *UP from the Ape*, New York, McMillan Company.

Howells, William (1973): *Evolution of the Genus Homo, Reading Mass*: Addison Wesley Publishing Company.

Hulton, J. (1975): *The Theory of Earth*, 2 Volumes, Edenburg.

Huxley, J. (1943): *Evolution: The Modern Synthesis*, New York: Harper and Row.

Huxley, Th. H. (1863): *Man's Place in Nature*. London.

Jain, K.C. (1979): *Prehistory and Protohistory of India*. New Delhi: Agamkala Prakashan.

Jonnes, F. Wood (1929): *Man's Place Among Mammals*, New York: Longmans Green.

Korn, N. and H.R. Smith (1959): *Human Evolution*, New York: Henry Holt and Company.

Kurten, B. (1972): *Not from the Apes*, New York: Vintage Books.

Montague, M.F.A. (1961): *An Introduction to Physical Anthropology*, Illionois, Cearless Thomas.

Oakley, K.P. (1958): *Man the Tool Maker* (4th Edition), London: British Museum.

Plenderlith, H.J. (1956): *The Conservation of Antiquities and Works of Arts*, London: Oxford University Press.

Pyddoke, Edward (1961): *Stratification for Archaeologists*, London: Phonexin House.

Reddy, V.R. (1987): *Elements of Prehistory*, New Delhi: Mittal Publications.

Sankalia, H.D. (1962): *Indian Archeology Today*, Pune: Decan College Press.

———, (1964): *Stone Age Tools Pune*: Decan College Press.

———, (1974): *Prehistory and Protohistory India and Palaistan*. Pune: Decan College Press.

Singh, M.M. (1961): *Puratatva ki Ruparekha, Patna*: Purajyoti Prakashan.

Trigger, Bruce, G. (1968): *Beyond History*: The Methods of Prehistory, New York: Holt, Rinehart and Winston.

Verma, R.K. (1977): *Bharitya Pragaitihasik Sanskritiyan*, Allahabad: Pramjyoti Prakashan.

Wheeler, R.E.M. (1956): *Archaeology from the Earth*, London: Pelicon Book.

Zeuuer, F.E. (1959): *The Pleistocene Period,* London: Hutchinson Scientific and Technical.

Index